Quirky, readable and honest. Jonas and Eddie
to bringing the EITI to this stage of maturity.
important and honest. But the most crucial lesson—which they underline—is humility.
**Clare Short, Chair of the EITI; former Secretary of State for International
Development**

The great progress towards transparency in the oil and gas sector is very
encouraging. In my memoir, *Beyond Business*, I wrote about some of the
early steps towards the EITI. Now Eddie and Jonas have written *Beyond Governments*, teasing out some lessons from the EITI. I am pleased that they are
telling the story of how change really happens.
**Lord Browne of Madingley, Chairman, L1 Energy; former Group Chief
Executive, BP**

It makes me proud to read Eddie and Jonas's book. It has been such a pleas-
ure working with them. The EITI is state-of-the-art moving further into nor-
mative territory every year. Transparency and the fight against corruption
require multi-stakeholder approaches. In this book they provide a framework
that I draw on as I continue the quest for transparency also in other sectors,
such as the garment or fisheries sectors. The multi-stakeholder approach
described so vividly in this book may come close to being the "silver bullet"
we are all looking for when struggling for better global governance.
**Professor Peter Eigen, former Chair of the EITI and Founder of
Transparency International**

In an era of widespread governance failures there is a growing need for collaboration. The EITI offers valuable insights. The experiences Jonas and
Eddie share are of great importance for all corporations that understand
that long-term financial success goes hand in hand with good governance
and good environmental and social stewardship.
Georg Kell, Executive Director of the UN Global Compact

Transparency is not only about fighting corruption. It is also about building trust. Collective governance is an underused tool that must be applied
along the extractive industries value chain.
Eleodoro Mayorga Alba, Former Minister of Mines in Peru

For investors, whether corporates sinking billions into decade-long projects, equity investors taking a stake in extractive companies, or sovereign bond investors betting on an emerging country's development, initiatives like the EITI have always been about driving up value by driving down country risk. As such, the business case of the EITI is to shift the risk–return equation by squeezing out opaque business practices, and allowing the discipline of transparent markets to bring about competitive, sustainable financial returns.

Karina Litvack, former Head of Governance & Sustainable Investment, F&C Asset Management

Bringing about change in my country required a collective determination towards openness and transparency—the power and locomotives of our development. The EITI contributed to providing this and this book helps others to do so elsewhere.

Roza Otunbayeva, former President of Kyrgyzstan

The EITI has come a long, remarkable way since our first discussions in 2002. Business today needs to come together with civil society to help governments govern better. This book provides important lessons from one of the more sophisticated efforts.

Mark Moody-Stuart, Chairman of Foundation for the United Nations Global Compact; former Chairman of the Royal Dutch/Shell Group and Anglo American

In Nigeria the EITI sits at the heart of contentious debate about the governance of our oil, gas and solid minerals. The debate still has a long way to go, but at least it is now informed by facts and not just innuendo, intrigue and allegation.

Ledum Mitee, Chair of Nigeria EITI and long-time Nigerian oil sector activist

My country, Timor-Leste, has been gifted with a once-in-a-lifetime opportunity, which comes in the form of oil and gas. The building of our nation hinges on the success of our governance. As a post-conflict nation, with human capacity and infrastructure challenges, a different form of governance is necessary. This book guides those going through similar journeys.

Alfredo Pires, Minister of Petroleum and Natural Resources, Timor-Leste

Development is more about good leadership and governance than it is about money. Looking back, I am very proud that Norway decided not just to support the EITI, but to do it at home. It has led to a global standard. We must get better at these collaborative governance efforts. Eddie and Jonas's book helps us on the way.

Erik Solheim, Chair of the OECD Development Assistance Committee (DAC); former Minister of Environment & International Development, Norway

The book is timely and fills a gap in the literature of practitioner guidance. Accessible, provocative and not afraid to challenge.

Michael Jarvis, Program Leader, Governance for Extractive Industries, World Bank

This is more than just the definitive account of the EITI, valuable to those seeking developmental benefits from mining, oil and gas. It is the most grounded explanation of what governance is, and ought to be, that any scholar or practitioner of multi-stakeholder decision-making is likely to encounter.

Dr Kathryn Sturman, Senior Research Fellow, Centre for Social Responsibility in Mining, The University of Queensland

Beyond Governments

Making Collective Governance Work: Lessons from the Extractive Industries Transparency Initiative

Eddie Rich & Jonas Moberg

BEYOND GOVERNMENTS

GOVERNMENTS

MAKING
COLLECTIVE
GOVERNANCE
WORK

LESSONS FROM THE
EXTRACTIVE
INDUSTRIES
TRANSPARENCY
INITIATIVE

Greenleaf
PUBLISHING

© 2015 Greenleaf Publishing Limited

Published by Greenleaf Publishing Limited
Aizlewood's Mill
Nursery Street
Sheffield S3 8GG
UK
www.greenleaf-publishing.com

Cover by Arianna Osti (ariannaosti.com)

Printed and bound by Printondemand-worldwide.com, UK

British Library Cataloguing in Publication Data:
    A catalogue record for this book is available from the British Library.

        ISBN-13: 978-1-78353-189-9 [hardback]
        ISBN-13: 978-1-78353-185-1 [paperback]
        ISBN-13: 978-1-78353-183-7 [PDF ebook]
        ISBN-13: 978-1-78353-188-2 [ePub ebook]

Contents

Part 1: Introduction

Part 2: A brief history of the EITI

Part 3: How to be a governance entrepreneur—
a framework for managing collective governance

Foreword

Clare Short
Chair of the Extractive Industries Transparency Initiative (2011–16)

This short book by Jonas Moberg and Eddie Rich, the Head and Deputy Head of the EITI, is a personal account of the development of the EITI over the last ten years and the lessons to be learned by other multi-stakeholder initiatives.

There are many multi-stakeholder initiatives operating in the international system, and the EITI is one of the strongest from the recent crop: the International Labour Organization is much older and strongly established. It is therefore useful to hear from the two people at the core of the development of the EITI, how they saw the story unfold and what advice they would pass on to others.

Their account is quirky, readable and honest. I do not agree with everything in the book; however, I do agree that building the EITI has been a significant journey and that it is important to try to capture lessons learned and share them with others building multi-stakeholder initiatives or working within the EITI.

The authors quite rightly advise humility in working to build multi-stakeholder cooperation. It is remarkable that the EITI has nearly 50 countries in membership and that the numbers have built up very rapidly in recent years. I do believe that the EITI has helped to build an expectation of transparency in a sector that has in the past been deeply opaque; and also that, by bringing together governments, companies and civil society at country and international level, the EITI has helped to build trust and a focus on the real issues at stake, which will help to build better governance in the sector. But I also share their worry that multi-stakeholder organizations can be mesmerized by setting up arrangements and processes that satisfy each of the partners but produce few results for the people who need to see real benefits from the better management of their resources. The authors stress that the nature of multi-stakeholder work involves managing conflict and constantly seeking compromise, and my own conclusion is that there is a danger that simply keeping going feels like a significant achievement whether not there is much benefit in the real world.

So, as the EITI moves through infancy and adolescence into maturity, it is important that it is less focused on its own processes and more on impact; less on the production of the EITI reports in order to fulfil EITI rules and more on improving government systems so that transparency and accountability are enhanced across the world.

Jonas and Eddie have contributed enormously to bringing the EITI to this stage of maturity. The lessons they point to are important and honest. But the most crucial lesson—which they underline—is humility. There is a lot yet to be done to ensure that, in the countries that are members of the EITI, the principles on which the movement was built—most particularly that natural resources should bring real benefits to improve the life of the people of resource-rich countries—become a reality.

Clare Short
April 2015

Preface

The road of development history is littered with the corpses of dead, well-intentioned initiatives that promised to solve global problems—Green Revolution, Big Aid, Washington Consensus, Budget Support, Public–Private Partnerships, Information Technology revolution, etc. Each fad has taught us a little more, but each has delivered less than promised. Transparency and big data, codes and standards, and collective governance appear to be in favour at the moment. Peer review, governance indices and continuous adaptation are making these ideas more robust. Like previous fads, they have much to commend them, but the enthusiasm of a school of academics needs to be tempered and adapted by the experience of everyday practitioners.

To read academic literature one would think that collective governance is common practice. Yet surprisingly few collective governance initiatives have been able to retain commitment or relevance. Collective governance practice, not just multi-stakeholder consultation but real collective decision-making, lags mournfully behind academic writing and the development debate. What practice there is has not been well documented, and consequently the challenges of process have been overlooked.

In practice, managing collective governance is a tough and slow process, not for the faint-hearted. It is a form of governance when regulation, legislation, taxation, nudging, consultation, enquiry and other more conventional government tools have been exhausted or are not viable. Although arguably one of the more successful, even the Extractive Industries Transparency Initiative (EITI) process has left bruises. This is to be expected—solutions to seemingly intractable governance problems are, by definition, difficult.

But there are some good practices for a difficult area, and we therefore wanted to document our experiences. If collective governance and multi-stakeholderism is to be part of modern governance, we must all get better at learning from what works and what does not. We have to be faster at sharing good practices to other efforts.

The authors are practitioners, not academics. At best we are governance entrepreneurs experimenting on the boundaries of government's capabilities. We hope that this book can be a useful guide for those facing similar challenges, whether it is in fighting corruption, promoting human rights, governing construction, medicines, arms, ports, environmental issues, urban planning, press regulation and other governance challenges of the 21st century.

This book is divided into four main parts:

1. An introduction, entitled "The irresistible rise of collective governance?", which seeks to explain the concept of collective governance and why it has become such an alluring concept in academic and development forums. Yet beyond the discussion, theory and acclaim, we conclude that there is limited practice.

2. A brief history of the EITI, "Collective governance in practice", which is the main experience of the authors and from which we draw our more general conclusions about governance entrepreneurship later in the book.

3. The main section of the book, "How to be a governance entrepreneur—a framework for managing collective governance", which serves as a manual for collective governance including the preconditions for the model, how to build trust through momentum, and governing the governance. This section also challenges the conventional wisdom on the mechanics of collective governance. Rather than seeking to agree long-term objectives among conflicting and diverse groups, a governance entrepreneur starts out seeking a narrow consensus from which trust and progress can develop.

4. Recommendations and conclusions, including the applicability of a collective governance model to other sectors, and recommendations for the international community.

"Beyond" governments should not be interpreted as "instead of" governments. Collective governance should reinforce and strengthen government systems and build the ability of the formal institutions to do their jobs. Indeed, the mark of success for such interventions, including the EITI, should be the ability to graduate out of the need for formal collective governance. However, at least for now, certain challenges cannot be addressed by governments alone. For example, in the governance of oil, gas and minerals, collaboration with civil society actors and companies is required for the benefits to be shared.

The title of this book, *Beyond Governments*, has been deliberately chosen to be seen as a compendium to other important books: *Beyond Business* (John Browne's memoirs which in itself was a play on his rebranding of BP as "Beyond Petroleum")[1] and *Beyond Charity* (Rockefeller Foundation).[2] Various areas of corporate social responsibility and corporate governance have also become known as "beyond compliance",[3] and we named the 2013 EITI global conference "Beyond Transparency".

The subtitle, "Making collective governance work", is explained in the subsequent chapters and is underpinned by the EITI as a case study of collective governance in practice. The challenges, lessons and opportunities of the process are presented with the intention of leaving the reader with some answers as to whether the collective governance model is as alluring as much of the literature would have it, and if so, in what circumstances.

The third and main part, "How to be a governance entrepreneur—a framework for managing collective governance", is intended to stand alone as a manual for managers of collective governance initiatives.

The authors' views are those of practitioners. The book is written from our personal experience, though reference is made to academic literature which we found to be both often illuminating and often misguided. Our conclusions not only sometimes run counter to the academic mainstream, but they are also, in places, counter-intuitive. It is, for example, suggested by the literature reviewed by us that collective governance efforts, like so many others, should start by identifying clear objectives, missions or goals. Not so, we suggest. Collective governance is ultimately a form of what in German can be described as realpolitik,[4] an effort searching for practical solutions,

---

1   Browne, 2010.
2   Abrahamson, 2013.
3   Prakash, 2001.
4   Realpolitik: "A system of politics or principles based on practical rather than moral or ideological considerations. Origin—early 20th century: from German

rather than based on an ideological ambition. When it comes to collective governance, it is likely to be destructive to seek agreement among hugely different stakeholders on a clearly defined ambition. Instead we recommend a process well known to mediators and relationship counsellors: start off by accepting that different actors may want different results from the process. Then, instead of seeking agreement on the end result, or even on the most important issue, it is necessary to seek consensus in any way that keeps the disparate groups around the table. From that platform, trust can be built, and the **consensus can move from the narrow to the meaningful**.

Realpolitik 'practical politics'" (http://www.oxforddictionaries.com). In this respect, it shares aspects of its philosophical approach with those of realism and pragmatism.

Acknowledgements

Our thanks go to Clare Short, Peter Eigen and all those that have served on the EITI Board, Sam Bartlett, Dyveke Rogan and other members of the International Secretariat, Rockefeller Foundation Bellagio Centre, all of the thousands of people who daily contribute to the implementation of the EITI, wives and family who have not chosen but graciously accepted to live with the EITI.

Sam Alston, Edward Bickham, Stuart Brooks, Richard Dion, Gavin Hayman, Michael Jarvis, Mark Moody Stuart and Pablo Valverde, who have reviewed various drafts of all or part of the book.

Thanks also to Rebecca Macklin at Greenleaf Publishing for her support and inputs at all points. Also thanks to Motoko Aizawa, Alexander Gillies, Dani Kauffman, Erica Westenberg and others who have been so helpful in various developments of this book.

Part 1: Introduction

1
The irresistible rise of collective governance?

1.1 What is collective governance? Much talk, less applicability, even less practice

As new models of development evolve, so does new vocabulary. Some of these are clumsy, unwieldy and ill-defined. For example, the lack of an agreed definition of collective governance has made theorization more difficult and more contested than it need be. In addition, we have added to the lexicon new expressions that seek to describe new concepts. All of these need explanation.

"Governance" is, according to the World Bank,[1] about the processes by which bargains between state and society are made, including policies and institutions, and how they are subsequently implemented and monitored by organizations. Related to this is "government", which encompasses the institutions which govern a society.

Experience and history points to some areas of governance that, either through weakness of enforcement, or inability or disincentive to legislate, have not been addressed by the institutions of government. Collective governance[2] has arisen as one of many responses to these governance gaps.

---

1   Barma *et al.*, 2012.
2   This term is used synonymously with "multi-stakeholder governance". We preferred the term "collective governance" because "stakeholders" can be a misunderstood notion, and to avoid confusion with "multi-sectoral". However, in EITI and other literature, "multi-stakeholder" is more commonly used. "Collaborative governance" is also used by some writers but we were concerned that

Other responses include negotiations of international treaties; international law to force governments to act; and soft law measures such as company self-regulation and public–private partnerships.

The term "collective action" is increasingly used in anticorruption contexts. It is based on the understanding that corruption can in at least the short term benefit the individual actor paying the bribes, even if the collective suffers in the long term. The solution to this Prisoner's Dilemma may lie in everyone coming together to act collectively. As long ago as 1965, Mancur Olson wrote in his seminal work *The Logic of Collective Action: Public Goods and the Theory of Groups*: "Indeed, unless the number of individuals in a group is quite small, or unless there is coercion or some other special device to make individuals act in their common interest, *rational self-interested individuals will not act to achieve their common or group interest*"[3] (his italics). We consider the EITI to be the kind of special device that makes individuals act in the common interest.

Building on this, the authors define collective governance as "the formal engagement of representatives of government, civil society and companies in decision-making and in public policy discussions". This definition draws on Ansell and Gash[4] and Donahue and Zeckhauser.[5] It has four key elements. The first is that the engagement is **formal**, i.e. it is somehow institutionalized. The government must establish the formal institution, since it is first among equals—collective governance without government is a

collaboration could be interpreted to include consultation models of governance. "Collective action" is also an increasingly used term, but the authors were keen to specify that the action was governance. "Shared value" is similarly more widely used, but this term appears to apply more to corporate social responsibility interventions than public policy discussions.

3   Olson, 1965.

4   Ansell and Gash (2008) provide six criteria for collective governance: the forum is initiated by public agencies; participants include non-state actors (citizen participation); participants engage directly in decision-making and are not merely "consulted" by public agencies; the forum is formally organized and meets collectively (continuous and institutionalized); the forum aims to make decisions by consensus (even if consensus is not achieved in practice); and the focus is on public policy or public management (public problem-solving).

5   Donahue and Zeckhauser (2011) also give six "dimensions" of what they call "collaborative governance": formality (formal to informal to tacit); permanency (permanent to ad hoc to temporary); focus (narrow to broad); diversity of participants (number, type); stability, defined in terms of shared objectives; and discretion, which is "the most useful discriminant for separating collaborative governance from other forms of public–private interaction".

contradiction. There are numerous collective action initiatives that do not include government membership. It is worth noting that many of these were created with a specific purpose to forestall the need for government action that would mandate certain actions. So, of course, governments are not a factor in those efforts also, but they are not strictly "collective governance" initiatives.

Formality is necessary in order to ensure that all bodies have a right to be at the table. In the EITI, the participants from government, companies and civil society are formally part of a self-governing forum. This has been a critical element for the protection of civil society in the EITI.

Second, that it includes **government, civil society and companies**. Governments and civil society are essential in order to ensure accountability. In theory, companies could be replaced by another relevant constituency (such as international development agencies or charitable foundations). Similarly, there might be more than three constituencies. The EITI Board includes "supporting countries" who are effectively donor agencies and political supporters, and who sit alongside the implementing governments, though caucus separately. It also includes institutional investors, who sit alongside the companies but have slightly different interests. Legislators, lawyers, industry associations, private foundations, academics, trade unions and faith groups might all be part of the collective set-up without a clear constituency among the three. In any case, it is essential for collective governance that there be at least three constituencies since it is necessary

The power of three

The case of the extractive sector is illustrative of the need to have government, companies and civil society around the same table. Civil society often has a suspicion that government and companies are in cahoots together to slice them out of the deal. At the same time, companies may conclude that governments and civil society are colluding to keep foreign companies out and or to renegotiate contracts. Finally, governments can get concerned that communities and civil society are blaming them for the lack of services arising from the taxes and royalties paid by companies. Consequently it is essential that all parties can *collectively* hold each other to account, rather than bilateral discussions where problems are blamed on those outside the room.

There might, of course, be other stakeholders to bring around the table. Parliamentarians, industry lobby groups or company foundations do not fit easily into any of these categories. The key is to bring the right players around the table, not to follow some strict formula of constituency labels.

for any one corner of the triangle to challenge the relationship between the other two.

Third, they must engage in **decision-making**: this distinguishes collective governance from the thousands of other "multi-stakeholder" consultation or advisory processes. Decisions are made collectively, often by consensus, possibly with any single stakeholder group being able to veto a decision. Having non-state actors formally as part of decision-making is in fact a surprisingly rare phenomenon. In the EITI, the national commissions in implementing countries decide what information needs to be collected to inform the debate on the extractive industry's challenges in that country.

The final element is that they must in engage in **public policy discussions**. Collective governance efforts such as the EITI seek to strengthen, not substitute, government institutions by bringing complementary expertise to provide policy advice. In the case of the EITI, the national commissions agree to publish information, possibly even analysis, and to discuss, though rarely agree, the implications. The final public policy decision remains with the government.

This book only seeks to explore this limited concept of collective governance as practised by the EITI, i.e. those efforts that focus on governments, companies and civil society.

The laudable Ethical Trading Initiative is not included as a collective governance effort as it is between companies, trade unions and civil society—there is no formal government involvement. This works well for the global supply chains of retail companies but is not the focus of this book. Similarly, the Forestry Stewardship Council (timber), Global Initiative Network (ICT) and 4C Association (coffee) do not seek to include governments in their governance structures.

Other frequently cited examples, such as the Kimberly Process Certification Scheme and the European Defense Integrity Initiative, are essentially intergovernmental processes with formal non-state observation rather than engagement in the governance. The UN Global Compact is a code proposed by the Secretary-General with multi-company signatories and civil and labour union support. It is formally endorsed by the UN General Assembly every two years. The Partnership Against Corruption Initiative and Transparency International's Integrity pacts are essentially multi-company codes of conduct with civil society—and occasionally government—participation, albeit often of a limited nature. Public–private partnerships are essentially contractual arrangements for service delivery. "Advisory groups" might be multi-stakeholder but do not have decision-making roles or are only bilateral, such as parent–teacher associations and community-based policing.

It is important to emphasize that this way of defining multi-stakeholder governance is not a value judgement by the authors. As the reader will see, under this restrictive definition we do not evangelize about the collective governance model as the solution to all governance gap problems, and we lack the expertise to suggest that it might be a better model for the issues that the above initiatives are trying to address. We suspect not. The point of the definition is to be clear from the beginning about what is being discussed.

The reader is also asked to forgive the authors for a bias towards international initiatives throughout the book. This is merely a reflection of our expertise, not that the approach should be exclusive to an international environment.

More generally, we hope that many of the lessons might also be applicable beyond our narrow scope of collective governance—it should obviously not be a case of only do this if your initiative ticks all the above boxes, but rather see whether this approach might apply to your challenges.

In any examination of these initiatives, it is quickly apparent that there is an over-representation of cases of natural resource management (not just oil, gas and mining, but also timber, coffee, agriculture, etc.). This says a great deal about the nature of the challenge of these sectors.[6]

The authors also introduce the term "governance entrepreneurship". This should not, of course, be confused with the concept of making money from governance. Rather it appeals to the more traditional entrepreneurship concept of **being creative**. Governance entrepreneurship can be defined as the inventive approach to solving intractable governance challenges through organization, evaluation, operation, experimentation and risk-taking, often in uncertain, hazardous and conflicting situations. The use of the term in this book applies almost exclusively to the conveners of the collective governance process. At times, other stakeholders are praised for their entrepreneurship and leadership, but that is not to be confused with the main convener role towards whom Part 3 of this book, "How to be a governance entrepreneur—a framework for managing collective governance", is focused.

6   Ansell and Gash, 2008.

1.2 The irresistible rise of collective governance?

> The idea of the future being different from the present is so repugnant
> to our conventional modes of thought and behaviour that we, most of
> us, offer a great resistance to acting on it in practice.
>
> John Maynard Keynes

1.2.1 A topology of turbulence: diplomacy is dying

> Prediction is very difficult, especially if it's about the future.
>
> Nils Bohr, Nobel laureate in physics

The case for collective governance is compelling. In a world of complex
policy challenges and shifting powers, there needs to be a space where the
main interests and experts can come together to solve policy problems. As
Sir Mark Moody-Stuart, former chair of Shell and AngloAmerican and long-
standing supporter of the EITI, writes: "It is only through some form of collective action involving the co-operation of other elements of society that
many issues can be addressed."[7]

The world used to be run by states. Where there were no strong state institutions but natural resources to be plundered, companies would take on
the mantle and form of statehood in collusion with the local elites. Daniel
Litvin[8] describes how the East India Company in the 18th and 19th centuries
interwove profits, politics, theft and self-deception. It conspired with the
Mughal political elite to profit from the people and the natural resources.
Later in the 19th and earlier 20th century, Cecil Rhodes did much the same
with his British South Africa Company in pursuit of gold and diamonds.
Joseph Conrad saw the same in the Congo from "sordid buccaneers" who
aimed "To tear treasure out of the bowels of the land ... with no more moral
purpose at the back of it than there is in burglars breaking into a safe". The
pattern was repeated by the United Fruit Company in backing a coup in
Guatemala in 1954 that set off civil strife for the next 50 years.

Where the states were too strong and ignored or were oblivious to the will
(or more often hunger) of their people, there might be "people revolutions".
In 1649 in England, 1789 in France and 1917 in Russia, civil society effectively took over the reins of government rather than offer a different form
of governance.

7   Moody-Stuart, 2014.
8   Litvin, 2013.

States are still powerful, but their power is different. The world is changing faster than at any point in our history. Obviously power is shifting from North to South and from West to East. But more importantly, now the world has four centres of power—governments, companies, finance and civil society—instead of a world dominated by governments. Governments everywhere are losing out to cross-border networks. This can be seen with transfer pricing tactics by companies, migration, international financial flows, climate change, land acquisition, terrorism and Internet technology. This means that governments have to deal as much or more with international governance institutions, with companies, with banks and with individuals, than with each other. Perhaps diplomacy between states is dying. Long live collaboration between state and non-state actors.

Ambassador Tormod Cappelen Endresen, a Norwegian diplomat who was instrumental in setting up the EITI International Secretariat in Oslo, coined the phrase "innovative diplomacy" when reflecting on the early evolution of the EITI. The authors understand this to be the approach to international relations and negotiations which recognizes the changing relations of states and non-state actors, and adjusts models of persuasion and incentivization skilfully to these realities. As with governance entrepreneurship, the phrase implies a willingness both to try new approaches and to take risks.[9]

1.2.2 Collective governance as a response to the race to the bottom

Governments are generally not well equipped to deal with many stakeholders on so many complex cross-border issues. In the 1990s, these transnational trends were considered to be leading inevitably to a "race to the bottom".[10] The only way states could attract investment was seen as by removing regulation or enforcement to allow companies to cut costs on labour rights, environmental safeguards, tax responsibilities, human rights, transparency, etc. In many cases, regulation could be bypassed through

9 Along with the EITI, our favourite example of innovative diplomacy is the Open Government Partnership (OGP): a nationally owned international process, a sort of marketplace for policy ideas driven by peer pressure between countries around some loose principles and codes. The policies to which the countries commit are then monitored by civil society and the international forum. Neither well funded nor defined, it is driven and incentivized by the top and by peers. Looking ahead, the OGP will find it hard to maintain momentum, as focus is now shifting from the content of OGP action plans to the implementation review mechanism (IRM).

10 Morrison and Wilde, 2007.

corrupt facilitation payments. The temptation for companies to do this was overwhelming where countries had weak institutions and processes. It was particularly acute in the case of extracting natural resources, where the company could not choose the government environment but needed to get to the resources ahead of its competitors.

Tempting though such behaviour was for companies, on the whole they recognized that they did not benefit in the long term from poorly regulated environments. Such environments were unpredictable, unstable and facilitation payments could be arbitrary and expensive, yet they were locked into the race to the bottom. They faced a Prisoner's Dilemma—they would all benefit from a different system, but any individual move could be grossly disadvantageous for that company.

Litvin continues in his book to describe the multinationals of the late 20th and early 21st centuries. According to him, their instincts for power might not so much have changed but been softened by the practicalities of doing business safely. He writes, for example, about Shell in Nigeria that "the storm over Saro-Wiwa forced a fundamental rethink by the company of its approach to social and environmental issues across the world, not just in Nigeria". De Beers came across similar challenges with blood diamonds; Nike with child labour. All need to respond to a growing disquiet about capitalism as encapsulated in, for example, the Occupy movement.

Self-regulation by companies around a set of agreed codes and standards, as with the Responsible Jewellery Council, became the initial response to the race to the bottom. It soon became apparent that, in international business, more was needed than company action alone. If competitors eventually managed to agree a code of conduct which contained any awkward provisions for government about, say, transparency or refusal to bribe, the governments could always find new actors. Furthermore, from an activist point of view, business self-regulation was of questionable effectiveness and legitimacy—who was enforcing, who was monitoring, wasn't collusion possible? Reformist governments saw the need to organize these efforts into formal "soft law"—a sort of nonbinding regulation which incentivizes good behaviour. This included legal provisions for more transparency in operations, contractual requirements to agree to codes of conduct, or high-entry requirements for bidding for tenders, such as Transparency International's Integrity Pacts.

Meanwhile, something else was happening. Anyone literate with a computer or smartphone—man, woman or child—was no longer locked out of politics. There will be a billion more middle-class people over the next 15 years: mostly educated, mostly politically aware and mostly demanding better lives than those of their parents. Social media was a new, more effective means for hungry and angry people to overthrow a government. This was evident in the Arab Spring, but it was not limited to the Middle East and North Africa. Several governments in Africa, Asia and Europe have been rocked by popular uprisings and many have radically shifted their policies as a response. Civil society, with its varied nature and international networks, is nimble at reacting to international changes. It is becoming stronger, and better at moving beyond advocacy to become part of problem-solving, not just problem identifying.[11] At its best, it can shine a torch for the direction the governance should take, both nationally and internationally. At its worst, it lacks accountability, can distort voices and interests, and can be susceptible to misguided and violent doctrine.

The multinational companies are, on the whole, adapting well to the opportunities of these massive new consumer markets, but there are also examples of how they have struggled to establish local roots and win over the trust of this new middle class and the emerging political power in countries in which they operate. They may be distrusted and some are still mired in much of the race-to-the-bottom behaviour.

There is an urgent need to reassess the comparative roles in a global society of governments with their mandate and power to regulate and provide public goods; companies as drivers of economic growth and employment; finance with its vast ability to move money and determine sustainable business models; and civil society emboldened by increased literacy, technological access and international links. Power has shifted from formal institutions, such as the state, to informal and nebulous processes, such as the Internet and social media. Power has also shifted from the state to the market. These shifts have jumbled up the old arrangements about who has a voice in society and, as in the Arab world, opens up the spectre of violence. The old models of nation-states do not adequately capture the complexity of governance in the 21st century.

This fragmented world is deeply unsettling and unpredictable for governments. They are still the arbiter of public policy but have to work differently to maintain their mandate. They have to learn to work with companies, finance and civil society differently. Soft law has temporarily stemmed, or at least slowed, the race to the bottom in many areas but, to establish the trust of their citizens, governments need more legitimacy. There are two,

11 Cramer, 2013.

often contradictory, approaches governments can adopt to overcome this governance gap, to reassert their governance role and to establish trust with the citizens. The first is to grab institutions away from other centres of power—for example, by bringing an ever larger share of natural resources into the hands of state-owned companies, also known as "resource nationalism". This is, at times, unfortunately combined with bringing the media under government control. The other approach is the more long and difficult route: engaging more thoroughly on governance issues with citizens and companies.

In 2007, in his report to the UN Human Rights Council, John Ruggie documented several cases of "a new multi-stakeholder form of soft law initiatives" which he saw as emerging:[12]

> These initiatives may be seen as still largely experimental expressions of an emerging practice of voluntary global administrative rulemaking and implementation, which exist in a number of areas where the intergovernmental system has not kept pace with rapid changes in social expectations ... The standard-setting role of soft law remains as important as ever to crystallize emerging norms in the international community. The increased focus on accountability in some intergovernmental arrangements, coupled with the innovations in soft law mechanisms that involve corporations directly in regulatory rulemaking and implementation, suggests increased state and corporate acknowledgment of evolving social expectations and recognition of the need to exercise shared responsibility.

So here was an extension of the soft law concept by which third-party civil society representatives act as monitors of the process and engage in discussions about the public policy implications.

But this approach was not instantly or instinctively straightforward for autocratic countries. Ian Bremmer's J curve[13] explains why resource nationalism is such an attractive option for addressing the governance gap for autocratic states. He starkly sets out the danger of moving from autocracy to a more open form of governance, arguing that it is in their transition that they experience the greatest instability. He notes that while many countries are stable because they are open (the United States, France, Japan), others

---

12   Ruggie, 2007. In the report he cites in particular the Voluntary Principles on Security and Human Rights, which promote corporate human rights risk assessments and the training of security providers in the extractives sector; the Kimberly Process Certificate Scheme, which seeks to stem the flow of conflict diamonds; and the EITI.

13   Bremmer, 2006.

are stable because they are closed (North Korea, Cuba, Iraq under Saddam Hussein). Regions of the world presently with the greatest instability are those in which the transition is happening or where the closed model has become unsustainable. When we look at the state of the Arab Spring countries, this appears to be a compelling notion.

For those closed states that, for whatever reason, do wish to open up, the implications of Bremmer's theory are that they will need significant and steadfast support for a "safe passage through the least stable segment of the J curve—however and whenever the slide toward instability comes". Domestically and internationally, it will mean engagement with citizens, companies and other stakeholders in new and unfamiliar ways. Transparency is also likely to play a role in making this passage from closed to open safer and faster. Capacity to engage in public policy discussions might be severally limited. There may well be areas, such as natural resource governance, in which Ruggie's extended soft law offers a contribution to this safe passage, and it seems that this is part of the reasoning for many countries to implement the EITI (see box on page 14 for the relationship between the EITI and the J curve). Many resource-rich countries, particularly in North Africa, are emerging from stable but autocratic political regimes to a new uncertain political future. Some of them see the EITI as offering them a route through this transition.

The oil, gas and mining sectors in a turbulent world

The oil, gas and mining sectors certainly raise a challenge that other sectors do not, in that countries do not compete in the same way for business. Instead, businesses compete for the opportunity to exploit the resource. As resources do not necessarily sit in well-governed countries, the sector is particularly susceptible to a corporate race to the bottom. And since the revenues are large, they can be particularly easily captured by a governing elite. These challenges disproportionately affect poor countries. These challenges can also be made worse by importing countries concerned about their own energy security and natural resource dependence.

In this topography of turbulence, where existing legislation and/or enforcement is not able to prevent a race to the bottom on corruption, human rights and other abuses, the rise of collective governance seems almost irresistible. It has emerged from the failure of autocracy or adversarial interest groups to govern against these issues. It was only one of many responses, but perhaps the one in which the gap between the acclaim and the practice is the widest.

The EITI and the J curve: transparency as a cure for instability?

Bremmer's J curve describes the relationship between a country's openness and its stability. The transparency process of the EITI in societies that are newly opening up might be a remedy to this J curve. There is normally considerable disruption from the release of data that might not have been easily available before. Stakeholders struggle "to handle the truth". The multi-stakeholder group acts as a sort of lightning rod for this disruption. From the experience of the authors, transparency has four potential positive impacts:

1. Diagnostic: the collection of data immediately helps to identify parts of the system that are not working. The institutions might not be collecting data or they might not be correctly recording it or, of even more concern, the institutions might not function or exist.

2. Deterrent: "sunlight is the best disinfectant"[14]–where there is record-keeping and transparency, there is less likely to be mismanagement and malfeasance.

3. Awareness-raising: information arising from transparency informs debate. Many become concerned that the information can be misleading or misused ("a little information can be a dangerous thing"), but even that will force stakeholders to explain themselves better.

4. Curative: transparency can identify mismanagement and determine resolution. It can then go further and inform and promote debate about better management of the whole system.

The EITI has a mixed record so far on these. With stronger efforts to improve the movement, there might be stronger impacts:

1. Diagnostic: this has been the EITI's trump card so far. The EITI has shone a light on where to improve tax- and data-collection systems and improved governance systems in every country which has produced a report. The Liberia report, for example, identified over 60 resource contracts and concessions across the extractives, forestry and agriculture sectors, revealing that US$8 billion of contracts were awarded in violation of Liberian law, i.e. as Leonard Cohen would say, "There is a crack in everything, that's how the light gets in."

2. Deterrent: it is difficult to know the EITI's impact on this, but there is lots of anecdotal evidence for changes in government agency and business practices.

3. Awareness-raising: the EITI has much further to go here. Information on process and availability of data is good, but context, analysis and meaning are still infant.

---

14   Brandeis, 1914.

4. Curative: the EITI doesn't have many stories on this. The government of Nigeria has recovered US$2 billion of unpaid taxes, the Democratic Republic of Congo around US$25 million, and Liberia has identified a case of company fraud. The EITI could be a great tool for making a more sophisticated assessment of a company's development footprint. How much is the impact? Is that a reasonable return given the investment—direct, and through taxation? Is the company paying enough and is it a good deal for the citizens?

Collective governance of transparency might therefore be part of the conditions most favourable for the regime's safe passage through the unstable segment of the J curve.

Part 2:
A brief history of the EITI

2
Collective governance in practice

The recognition that some complex governance challenges are best managed through collective approaches is fundamental to the EITI. To draw lessons on collective governance from the EITI, it is first necessary to chart the brief history of the EITI, describing what it is, how far it has come and why it has developed as it has.

2.1 Publish What You Pay and the launch of the EITI

In the late 1990s and early 2000s, there was an expanding library of academic literature around the "resource curse"[1] by such acolytes as Jeffrey Sachs, Joseph Stiglitz, Terry Lynn Karl and Paul Collier, detailing how the huge potential benefits of oil, gas and mining were not being realized and were associated with increased poverty, conflict and corruption. The problem went beyond just the well-known economic phenomenon of Dutch Disease, by which natural resource wealth made other export sectors uncompetitive and domestic sectors lost out to cheaper imports. Other common effects surrounded the capturing of the revenues by elites, corruption, the stunting of the development of tax systems to capture revenue from non-extractive sectors, and exacerbated regional and community tensions. These writings outlined the complexities of the governance of extractive resources—bidding, exploration, licences, contracts, operations, revenues, supply chains, state involvement, trading, local content, transit,

[1] Karl, 1997; Collier and Hoeffler, 1998; Bannon and Collier, 2003; LeBillion, 2003; Boschini *et al.*, 2007; Humphreys *et al.*, 2007.

services, allocations and spending. They noted environmental, social and political concerns. They each outlined remedies for addressing the curse, often noting that no single action would be capable of tackling all these challenges. However, the literature was clear—transparency and dialogue had to be part of the starting point.

These academic analyses were followed by more and more journalistic pieces[2] and a growing campaign by Global Witness, Human Rights Watch,[3] Oxfam America[4] other civil society organizations. International financier George Soros established a "Revenue Watch" programme under his Open Society Initiative, to investigate the flow of funds from oil companies to governments in the Caspian region. Non-governmental organizations (NGOs) stepped up their enforcement of corporate social responsibility rhetoric and were looking for a law for companies to report their payments to developing countries. The civil society campaign slogan of "Publish What You Pay" (PWYP) was drawn from a Global Witness report, *A Crude Awakening*. Launched in December 1999, it focused on the opaque mismanagement of oil in Angola. The report concluded by calling on the operating companies to adopt "a policy of full transparency [in] Angola and in other countries with similar problems of lack of transparency and government accountability".[5]

Responding to the campaign in February 2001, BP published the signature bonus of US$111 million it paid to the Angolan government for an offshore licence. It committed to publish more. This sparked a strong reaction from Angola. In his 2010 memoir, *Beyond Business*, Lord John Browne, the then chief executive officer of BP, recalled how he received a cold letter from the head of the Angolan national oil company, Sonangol, stating that, "[I]t was with great surprise, and some disbelief, that we found out through the press that your company has been disclosing information about oil-related activities in Angola".[6] The backlash and threats from the Angolan government led Lord Browne to conclude, "Clearly a unilateral approach, where

---

2   Roberts, 2006; Margonelli, 2007; Ghazvinian, 2008; Shaxson, 2008; Maas, 2009; Clarke, 2010.

3   Bronwen (1999) argues for greater transparency in government–company relations to "publish all documents related to payments, gifts, or contracts in relation to operations in the oil producing communities".

4   Ross (2001) recommended that host governments should provide "disclosures about all revenues they receive from extractive firms".

5   Global Witness, 1999: 3.

6   Browne, 2010.

one company or one country was under pressure to 'publish what you pay' was not workable".

The oil companies argued for a shift away from company reporting, as sought by PWYP and others, to reporting by governments, in order to reduce conflict with host governments and risk to contracts. If company reporting was to be required they wanted a global effort to level the playing field that required all companies in a country to disclose. Mining companies at this point became supportive, seeing the process as an opportunity to create a better understanding of their broader contribution to the economy.

The UK government—the Cabinet Office, the Department for International Development (at that time led by the future EITI chair, Clare Short), the Treasury, the Foreign Office and the Department of Trade and Industry—was listening both to the Publish What You Pay campaign and to the oil companies. They saw the opportunity to develop an initiative built on the notion of equal transparency from the governments and the companies.

The EITI is often thought to have been launched in 2002. It is true that the then UK prime minister, Tony Blair, outlined the idea of the EITI in a speech intended for the World Summit on Sustainable Development in Johannesburg in September 2002. However, the problematic relationship between Prime Minister Blair and President Robert Mugabe of Zimbabwe meant that the British prime minister never actually delivered his prepared remarks as intended.

2.2 Bringing stakeholders to the table: agreeing the EITI Principles

Following the publication of the Blair speech, the UK Department for International Development (DFID) convened a meeting of civil society, company and government representatives. There was agreement that some kind of reporting standard should be jointly developed. At a conference in London in June 2003,[7] a Statement of Principles to increase transparency of payments and revenues in the extractive sector was agreed.[8] These 12 EITI Principles centred on the need for prudent management of natural resources

---

7   Attended by "140 delegates representing 70 governments, companies, and industry groups, international organisations, investors and NGOs" (DFID, 2003).

8   EITI, 2003a.

based on enhanced information and public debate. They affirmed that there was a belief that "a workable approach to the disclosure of payments and revenues is required, which is simple to undertake and use".[9] Over 40 institutional investors signed a statement of support for the EITI which argued that information disclosure would improve corporate governance and reduce risk.[10]

Following this meeting, a few countries, such as Nigeria, Azerbaijan, Ghana and the Kyrgyz Republic, explored how these principles might be applied. They were later joined by Peru, the Republic of Congo, Sao Tome e Principe, Timor Leste and Trinidad and Tobago.

Transparency in natural resource development was championed at G8 summits in 2003 in Evian, and in 2004 in Sea Island, Georgia. The G8 subsequently called on the International Monetary Fund and the World Bank to provide technical support to governments wishing to adopt transparency policies. This led to the establishment of the World Bank-administered Multi-Donor Trust Fund (MDTF) for the EITI in 2004. With contributions from 14 donor countries, the MDTF has since disbursed almost US$70 million in technical and financial assistance to EITI programmes in almost 60 countries.

However, the resistance to implementation by the host country, the UK, was a problem. As Edward Bickham, then of AngloAmerican, told the authors: "it would be fair to note that the UK's non-implementation of EITI had become a political problem—if only at the level of debating points—for some African countries who saw it as confirmation of a neo-Imperialist mindset".[11] Eventually Norway committed to implement the EITI in 2006, paving the way for well-governed OECD countries to walk the talk. The EITI was making progress towards not just being a club of countries with governance challenges, but a global standard for all. The fact that OECD countries started to implement also made it politically possible to engage some large growing economies that hitherto had been disinterested, such as Brazil.

9   EITI Principle 10 (EITI, 2003a).

10   EITI, 2003b.

11   It is pleasing to note that the UK did eventually sign up to implement the EITI, officially becoming a candidate country in October 2014.

2.3 Drawing from countries' first experiences with EITI: the EITI Criteria

In March 2005, the EITI stakeholders and implementing countries again met in London for the second conference. The UK secretary of state for international development, Hilary Benn, summarized:

> Our experience in the four countries that have piloted EITI … is that while different countries have taken different approaches to implementation, this needs to be backed up by clear international rules of the game for the initiative to be effective and credible.

These different approaches to the principles were boiled down to six EITI Criteria, which sought to establish "the rules of the game". Benn also announced the establishment of an International Advisory Group, which would include representatives of governments, companies and civil society organizations, to take the EITI forward.

It became increasingly clear that the EITI was not evolving, as some had anticipated, into a voluntary corporate social responsibility standard for companies, but rather into a disclosure standard implemented by countries. The criteria focused on:

> Regular publication of all material oil, gas and mining payments by companies to governments ("payments") and all material revenues received by governments from oil, gas and mining companies ("revenues") to a wide audience in a publicly accessible, comprehensive and comprehensible manner.[12]

They also recognized that civil society had to be actively engaged in the process to enhance accountability.[13]

2.4 Assessing transparency: EITI validation

By the time of the third EITI global conference in Oslo in October 2006, the implementing countries (now joined by Niger and Cameroon) were preparing their first EITI reconciliation reports. Azerbaijan had already produced reports covering revenue from 2003 to 2005 and Nigeria a report covering 1999–2004. Alongside the production of an EITI Source Book in 2005, which

12   EITI Criterion 1 (EITA, 2005).
13   EITI Criterion 5 (EITA, 2005).

provided guidance on how to produce these reports, the International Advisory Group had sufficient emerging approaches to introduce the *EITI Validation Guide*, which set out the indicators that implementing countries had to meet in order to become EITI compliant.[14] The guide was introduced at the Oslo conference, effectively marking the end of the beginning for the EITI. The guide also included for the first time a formal process to sign up to become an EITI "candidate" country.

From 2002 to 2006, the EITI had been valiantly run by a small team in DFID, with little administrative capacity.[15] It was also agreed at this time that the EITI should have its own governance structure in order to grow and enhance credibility: a board, secretariat and a members' conference every two years to appoint the board. Peter Eigen, co-founder and former chair of Transparency International and hitherto the chair of the EITI's Advisory Group, was appointed as the first chair of the board. The EITI International Secretariat was later established in Oslo in September 2007 with Jonas Moberg appointed at its head. After two terms of office, Peter stepped down as chair and the Rt Hon. Clare Short was appointed as chair in Paris at the fifth EITI global conference in March 2011.

With the principles setting out its aims, the criteria containing its minimum requirements and the guide establishing its indicators, it was thought that the EITI had a structure in place that would clearly frame the expectations of implementing countries. The EITI, in effect, had evolved into a collective governance standard. In February 2009, Azerbaijan became the first country to be compliant with this standard, and was soon followed by Liberia, Timor Leste, Nigeria and Ghana.

2.5 Making the EITI more meaningful: the EITI Standard

Rules often provide a firm foundation but need to evolve according to experience and circumstance. It quickly became clear that the EITI Rules had left

14   EITI, 2006.
15   The EITI story is not complete without special reference to the civil servants in DFID who kept the process alive during these years of variable political support. They included, in alphabetical order, Tim Ayers, Malaika Culverwell, Sefton Darby, Mike Ellis, Daniel Graymore, Zoe Hensby, Mary Hunt, Ben Mellor, Diana Melrose, Simon Ray and Penny Williams.

many issues open, such as how long implementing countries had to meet the standard and how regular and timely the reporting needed to be.

So, in 2009 (fourth conference in Doha) and 2011 (fifth conference in Paris), the EITI Board issued versions of the EITI Rules.[16] Replacing the *EITI Validation Guide*, these included six "policy notes" that provided further clarification and guidance. The "indicators" became "requirements" and were addressed more as steps to be followed by implementers than as indicators to be assessed by external validators. The 2011 edition of the EITI Rules, for the first time, crucially included the need for the data to be both timely and regular.

Shortly after the 2011 Paris conference, an evaluation of the EITI by Scanteam was published.[17] The evaluation recognized exciting innovations from many of the implementing countries—for example, Liberia had included forestry and agriculture; Nigeria's reports included physical and process audits, as well as financial audits; Ghana's and Peru's reports included data on the amounts paid to subnational levels of government, etc. However, having consulted a wide range of EITI stakeholders, Scanteam concluded that "little impact at the societal level can be discerned ... largely due to [EITI's] lack of links with larger public sector reform processes and institutions".[18] It found that EITI's narrow focus was not *systematically* delivering on the principles established in 2003. Expectations for improved transparency were also growing. The EITI had to evolve to continue to offer leadership. The consensus could and had to be pushed further. The board and other stakeholders recognized that the EITI needed to do more to encourage countries to use the EITI as a platform for wider improvement of natural resource management.

The board undertook an extensive strategy review to address three main challenges:

1. How to ensure that the EITI provided more intelligible, comprehensive and reliable information

2. How to ground the process in a national dialogue about natural resource governance, i.e. linking the EITI with wider government processes around tax collection, extractive policy and budget arrangements

16   EITI, 2009.
17   Scanteam, 2011.
18   Scanteam, 2001: 1.

3.   How to incentivize continuous progress beyond compliance

Civil society naturally came to the table with a strong agenda, particularly around using the process to show better not just how much had been paid, but how much should have been paid—i.e. what value had been created to the government from the natural resources. This lead to many suggestions, including a focused demand for project-by-project reporting and contract transparency. For their part, companies were keen to see the process opened up to a greater scrutiny of budgets and transparency of expenditure choices; subnational flows; and accounting for barter arrangements. Beneficial ownership—i.e. the disclosure of the "flesh and blood" owner of each company—was championed by both sides, but did not end up as a requirement since there was little knowledge about whether countries were administratively ready to implement this. It is presently subject to a pilot to look into the possibilities.

The resulting EITI Standard,[19] launched at the global conference in Sydney in May 2013, therefore sought to:

- **Make EITI reports more understandable**: EITI reports were required to contain "contextual information" to underpin the data on revenues and payments, such as the contribution of the extractive sector to the economy, production data, a description of the fiscal regime, an overview of relevant laws, a description of how extractive industry revenues are recorded in national budgets, an overview of licences and licence holders, and a description of the role of state-owned companies. Countries were encouraged to publish contracts and details of the beneficial owners of companies.

- **Making EITI more relevant in each country**: countries were required to agree a work plan with objectives that articulated what they wanted to achieve with the EITI and set out how they wanted to achieve it. The scope of EITI implementation and links to other reforms had to be tailored to contribute to these desired objectives.

- **Better and more accurate disclosure**: the standard required for the first time that EITI reports disclose the payments broken down by company, by revenue stream and, in due course, by project. EITI reports were also to be made available electronically and codified to allow for international comparisons.

19   EITI, 2013a.

- **Recognizing countries that go beyond the minimum**: the standard introduced more frequent and nuanced validations to create incentives for more innovative use of EITI to the benefit of the country.

- **A clearer set of rules**: the EITI Standard was restructured from the EITI Rules, in order to condense the previous 21 requirements and policy notes into a shorter and more concise seven requirements.

By the publication of this book, the EITI had become a *global* standard. Almost 50 countries[20] were implementing the EITI, with France, Germany, Lebanon, Mexico and the Netherlands, among others, preparing to begin implementation. Over 200 EITI reports had been published covering well over a trillion US dollars of revenues paid. While the International Secretariat has remained relatively small, with a staff of around 20, over 400 people were working around the world on implementing the EITI and in excess of 800 people served on national EITI commissions.

In addition, various international institutions routinely cited their association with the EITI as evidence of their own commitment to good governance. The EITI's reporting requirements were reflected and, in some ways, exceeded in US, European, Nigerian and Liberian legislation, the World Bank's International Finance Corporation's standards for extractive projects, and an increasing set of country-level policies such as the publication of contracts and beneficial ownership.

Before 2013, the authors were worried that the EITI was going to become irrelevant by simply focusing on revenue transparency when the debate had moved on. By 2015, the concern was that everyone was trying to hang everything on it, because the EITI was the main game in town. In many countries, it was beginning to play host to some topics that had previously been considered politically taboo: beneficial ownership, production and consumer subsidies, commodity trading, the role and behaviour of state-owned companies, secretive contracts, aggressive transfer pricing, non-payment of taxes, smuggling, fraud, etc. The debate had clearly shifted and transparency was no longer an aspiration. It was an expectation. And through collective governance, it was beginning to lead to accountability.

20   For a full list of EITI implementing countries, see https://eiti.org/countries.

How the EITI works

The 2005 criteria in essence established core two elements:

Companies publish what they pay and governments publish what they receive in an EITI Report.

COMPANIES
disclose payments

EITI REPORT
where the tax and royalty payments are independently verified and reconciled

GOVERNMENTS
disclose receipt of payments

This process is overseen by a multi-stakeholder group of governments, companies and civil society.

COMPANIES · CIVIL SOCIETY

GOVERNMENTS
MULTI-STAKEHOLDER GROUP

Once these two elements were completed, the country EITI process was then "validated" to see if it had met the 21 EITI indicators (later requirements) that were established in 2006. Based on the report of the validator, if the board found that the country had met the indicators, it was declared "compliant". If it had not, it continued to be a "candidate", unless it was found to have made "no meaningful progress" in which case, it would be delisted.

With the 2013 standard, the scope of the process expanded as follows:

What is the EITI?

Ensures more transparency and accountability in the **natural resource value chain**

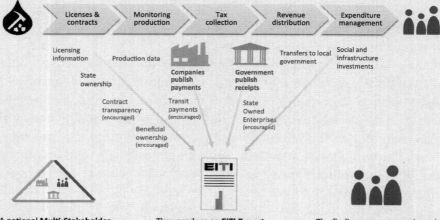

Licenses & contracts · Monitoring production · Tax collection · Revenue distribution · Expenditure management

Licensing information · Production data

State ownership

Contract transparency (encouraged)

Beneficial ownership (encouraged)

Companies publish payments

Transit payments (encouraged)

Government publish receipts

State Owned Enterprises (encouraged)

Transfers to local government

Social and infrastructure investments

A national **Multi-Stakeholder Group** (government, industry & civil society) decides together how EITI should work in their country and oversees the entire process.

They produce an **EITI Report** where government receipts and companies' payments are independently verified and reconciled.

The findings are **communicated** to create public awareness and debate about how they country should manage and use their resources better.

The core elements of revenue transparency from companies and from government remained, but there were also requirements from other parts of the value chain ranging from licences and contracts, to production, revenue distribution and expenditure management. Other elements were encouraged. There remained a strong focus on country ownership, with the national multi-stakeholder groups deciding the areas for data collection that they considered most relevant and meaningful for their process.

2.6 The EITI from here

The EITI is a Standard, but no standard should be an end in itself. Indeed, ultimately the EITI standard and process should be so embedded in government and company processes that it ceases to exist as a separate activity. The aim of the EITI should be to strengthen government (and company) systems, and inform public debate and trust building. To this end, the authors note three priorities for the future:

1. **To shift the focus from the *disclosure* to the *use* of data**: there needs to be greater effort to ensure that the data is reflected and used fairly, accurately and comprehensively, to inform public debate. A global database on the extractives encompassing EITI and other tax, spending, costs, prices, production, economic, social and environmental data would be a big step forward in improving both reliability and analysis. National databases on production, contracts, tax, royalties, social payments, etc., linked to existing mining cadastres and oil-gas concession maps, can help bring information into one easy-to-use place which would be interoperable with other data such as population, education and health services, roads, and so on. These things can only happen by linking with other data efforts in the field and ensuring that there is better codification of data. At the national and community level, it will require more emphasis than ever on explaining the data through "info-mediaries", "data tellers", data visualization, and so on. For companies (and indeed banks and institutional investors), the EITI could increasingly become a useful space to develop collective policy and guidance about tricky issues such as due diligence on buying assets. For example, how can one company have more information on the historical ownership of another company that it seeks to buy. For civil society organizations and others that have long campaigned for greater transparency, this shift in focus is a welcome challenge. It is a shift in part as a result of their efforts. While there is a long way to go, the EITI and many other complementary efforts are also leading to improved transparency and openness. However, it runs the risk of being a false victory unless these organizations are able not only to campaign for more, but also to make use of what there is. They will to a greater extent have to analyse available information and translate it into reform programmes. Campaigning for transparency has been successful and must increasingly result in, and be overtaken by, campaigning for better governance.

2. **To improve performance beyond compliance**: more than half the EITI implementing countries were already compliant by mid-2013. By 2015, it was approaching two-thirds. The standard evolved and would continue to do so, but essentially, like a driving test, countries would keep passing unless they crashed the car. A binary pass/fail is too crude to improve performance beyond compliance. There are many ad hoc communication mechanisms available to incentivize and recognize good performance, but there is also a greater opportunity to link EITI performance to the advancement of indices and ratings, such as the Resource Governance Index of the Natural Resource Governance Institute (formerly the Revenue Watch Institute),[21] the World Bank's Public Expenditure and Financial Accountability[22] and Country Policy and Institutional assessments,[23] the IMF's reports on the observance of standards and codes,[24] and sovereign debt ratings. The EITI is likely to have to reinvent itself in order to remain relevant, becoming more focused on promoting good performance than assessing technicalities (see Chapter 8).

3. **To link EITI implementation to international and national reform efforts**: recent years have seen the rise of a new type of development process—international marketplaces for national initiatives as typified by the Open Government Partnership.[25] The EITI will need to continue to piggyback on these processes to engender more peer learning, information sharing and co-ordination, and best practice. At the national level, the EITI processes need increasingly to act merely as portals and commentaries for strengthening data processes within government and companies, and accountability mechanisms within the existing governance framework—parliament, media, judiciary, etc. One could easily envisage EITI countries graduating from annual reports, instead providing a forum for the use of existing data and holding the government and companies to account.

21 Natural Resource Governance Institute, 2014.
22 See www.pefa.org.
23 World Bank, 2011.
24 IMF, 2014.
25 Open Government Partnership, 2015.

The EITI is in transition. The post-compliance agenda frees the International Secretariat from prescribers of the EITI Standard to facilitators of what countries want from the EITI. As one civil society voice in Tanzania said recently: "This is the discussion that we've been waiting to have." To paraphrase President Kennedy—ask not what you can do for the EITI, but what EITI can do for you.

However, many stakeholders—internationally and nationally—still have a tendency to get stuck in technical debates about discrepancies, materiality, coverage, auditing and, of course, **compliance**. EITI compliance is its currency. Having built Rome, it cannot be sacked in a day. Compliance was an important foundation for the wider debate, but can now be somewhat of a distraction. Compliance with the standard should not be seen as an endpoint in itself.

The EITI has since updated its validation process and made some good improvements, but the following questions that arose in the review of the validation process linger and raise significant conceptual challenges for other standards-based governance efforts:

- Should assessment of implementation only focus on whether all the EITI requirements are met? What about those elements that are "encouraged" rather than "required"?

- Should effort and progress in meeting the requirements over time be taken into account?

- Should the EITI recognize achievements that go beyond the minimum requirements?

- What emphasis, if any, should the validation place on progress with achieving national objectives, and the outcomes and impact of EITI implementation?

- To what extent should EITI assessment reflect progress in integrating EITI reporting in government systems?

- Should EITI requirements continue to be judged "met" or "unmet"? Or should there be nuance and possibly greater focus on narrative and written comments on the degree to which they have been met?

- Should the concept of compliance be replaced and, if so, with what?

- How can the EITI incentivize continuous progress and innovative implementation in compliant countries?

- To what extent should the EITI allow for more flexibility in how countries implement the EITI Standard, the approach adopted and the time needed?

EITI: the name

The name "Extractive Industries Transparency Initiative" has always been a mouthful. In many ways it is unsatisfactory. This story shows that it has ceased to be an "initiative"—it is now at least a standard, and increasingly more. "Transparency" is only half the story, for the process seeks to embed accountability and reform. "Extractive industries" is a clumsy term for "oil, gas and mining". The board could change the name. Yet it is an accepted brand, with some 50 implementing countries around the world having adopted the EITI brand. Just about all implementing countries have nationalized the EITI logo, i.e. taken the EITI logo and added the country's flag's colour or otherwise given it ownership. The name no longer belongs to the EITI Board alone. It works well in other languages—French, Arabic, Russian, Spanish and Portuguese. Instead, the International Secretariat has sought to reduce the brand to its letters, just as British Petroleum and British Gas first became acronyms and then eventually the names BP and BG. The secretariat has also developed and increasingly made use of tag lines such as "Transparency Counts" (2011–12), and "Beyond Transparency" (2013). The debate continues.

2.7 Conclusion

In little over a decade, the EITI has developed from a vague initiative into a multi-country, multi-stakeholder forum and then a global rules-based transparency standard, leading to an accountability process with minimum requirements. Each stage represented an important step in the progression. This experience with the EITI is another reminder of the importance of collective approaches to governance. It required civil society organizations to campaign as well as to engage; it required company leadership; it required representatives from supporting countries such as the UK to provide facilitation; and finally and arguably most importantly, it required leaders from implementing countries to respond and take ownership.

It is a story of a moving consensus; learning and adaption; increasing confidence; and strong leadership. Most of all it has been a story of **governance entrepreneurship**. These are the lessons for all collective governance processes which will be explored more in the following chapters.

Part 3:
How to be a governance entrepreneur—
a framework for managing collective governance

This part of the book establishes a framework for managing collective governance. If, as explained in Part 1, so many of the world's challenges are not solvable by governments, companies or civil society alone, why do so many partnerships fail? The International Civil Society Centre (ICSC) in Berlin undertook research which shows that, of the 330 partnerships in the study, only 24% could be described as a success against their self-reported function. A further 38% did not do anything other than have a launch, 26% had a few irrelevant activities and then petered out, and 12% had activities which led to some of their stated outputs.[1]

The techniques for governing collective governance efforts are quite different from those used to govern civil society organizations, multilateral organizations or corporations individually. Because collective governance is built on conflict, and because tensions come erratically and from all sides, it makes the process difficult to steer. Managing almost chaotic and dynamic interrelations takes a lot of enterprise and adaptation—a process the authors call "governance entrepreneurship".

Effective governance entrepreneurship requires:

- The analytical skills to recognize if the preconditions for collective governance exist

- The strategic skills to then build trust *through* building momentum

- The tactical skills to move the consensus from the narrow to the meaningful

- The interpersonal and diplomatic skills to win friends and promote leadership

- The institutional skills to "govern the governance"

- The wisdom to know when to close the institutions

These are the main pillars of "how to be a governance entrepreneur" and each will be explored in the coming chapters of Part 3, which draw unashamedly on the authors' experience with the EITI.[2]

---

1   Biermann *et al.*, 2012.
2   Morrison and Wilde (2007) undertook a survey of multi-stakeholder initiatives in the oil and gas sector with the following recommendations for successful implementation: a clear understanding of expectations regarding direct and indirect impacts; the clear promotion of the standard; the evolution of the governance structure; the development of practical guidance to aid implementation; the establishment of multi-stakeholder collaboration at implementation

level; a strong secretariat with independence and the ability to mobilize all stakeholders; the continuing political support of both home countries and producer countries; and a continuing ethos of leadership. From the experience of the EITI, this is a good list. This book will cover all of these issues, and specifically will dispute the first.

3
Preconditions for collective governance

A governance gap should not automatically mean a collective governance response. Just because collective governance solutions are increasingly popular, it does not it mean that they are increasingly right. They are not an end in themselves. Great society transitions do not tend to happen through alliances: they happen through better governance by governments. Collective governance should therefore only be a mechanism for strengthening governments. **Collective governance is an expression of crisis**: it is borne from the reality that existing models are not sufficient to meet the challenge.

There are two conditions that must exist before any consideration of taking the collective governance route.

3.1 Government failure: mind the governance gap

It may seem tautological, but it is important to recognize that if there is a government solution, then government should be the solution. Collective governance efforts should only come into play if there is an actual government failure, such as the inability to pass or enforce a policy, a regulation or legislation. This is especially likely in contentious policy areas such as corruption, human rights abuses or press freedom. In these areas, there are typically conflicts of interest for the governing elite who might be disproportionately open to influence by companies and their commercial interests. If the government is seeking to address these conflicts of interest, it can start by designing a code of conduct and/or providing more information, to create an accountability mechanism that addresses the governance gap.

Even if the government decides to develop a code of conduct or issue verified information to inform, there are many options that need to be considered before exploring whether to develop a collective governance approach. The government may, for example, consider consulting more widely, establish advisory groups or commission third-party monitoring. All of these options stop short of a collective governance solution.

It is also important that the problem requires long-term co-operation. Collaboration is much easier if it is not seen as a winner-takes-all one-off deal, but as part of an ongoing co-operation. This rules out mediation and conciliation around a specific one-off deal as suitable for collective governance.

Given this, it is no surprise that so much of collective governance revolves around information collection and around codes and standards for governments and civil society. These relate to behaviours that, even if they are enacted in legislation, need third-party monitoring to ensure, first, that they are reliably being implemented and, second, that they are informing and improving public policy.

The other area in which such an approach might be fruitful is around shared modelling, such as forecasting the costs of infrastructure or the future revenue from commodities. Where there might be lack of trust in the government and companies' planning or forecasting figures due to potential conflicts of interest, the results of the model could be made public or an independent model developed. Then representatives from government, company and civil society might come together to assess the independence of the model and the implications for public policy decisions.

The extractive sector and government failure

Angola has faced challenges that are extreme but common versions of many resource-rich countries. By the time the cold war was over, the economy became dominated by diamonds and, increasingly, oil and gas. Broadly speaking one side (UNITA) had the diamonds and the other (the MPLA government) had the oil and gas. The civil war continued until 2002, no longer funded by the old superpowers but by revenue from mineral resources.

Extractive companies set up in Angola with high capital investment sought long-term deals. Massive sums of money were passed between the companies and the government in "signature bonuses",[1] taxes and royalties. These became the main source of revenue for the government.

Furthermore, the massive investment from the one sector drove up the foreign exchange and priced out other exports, such as agriculture, making the extractives sector even more dominant—this is the so-called Dutch Disease, named after the decline of the manufacturing sector in the Netherlands following the discovery of a large natural gas field in 1959. As a result, the government relied more and more on oil and gas for its income. As the reliance on the extractive sector increased, so the reliance on other forms of taxation fell to the extent that the government found that it was barely worthwhile collecting taxes from people or businesses. All this weakened the social contract—with no taxes coming from ordinary people, their claims for public services fell increasingly on deaf ears. It is believed that much of the revenue from natural resources, which after all belonged to all the Angolan people, went into the pockets of the government elites and fuelled the conflict and repression.

The capture of revenue from natural resources by ruling elites is an extreme form of government failure. Big money or interests distort the political discourse and create conflicts of interest that need to be remedied, even in countries with strong institutions. For example, in the United States, the citizens could make little judgement about the fairness of revenue received from the mining sector because production figures were not centrally collected or disclosed. Such information is not required by government since, under mining legislation developed in 1876 during the gold rush, royalties are not collected—no royalties means no need for production information. Furthermore, mine-affected communities know little about the payments that are made between companies and district and chiefdom levels of governments. As with citizens in "resource cursed" countries, information and transparency was not going to be enough—citizens needed a seat at the table.

3.2 Conflict as a fuel: something to gain and a lot to lose

Anger, perhaps drawing on investigative reporting and campaigning, leads to a movement which leads to a need for resolution. Perpetual conflict is the fuel that keeps the wheels of collective governance moving. Governments,

---

[1]   The signature bonus system is common in many oil-producing countries. A payment is made upfront by a company to the host country for the right to develop a block commercially before work begins. It does not imply future revenues for the company from oil production in the same licence.

companies and civil society are organizations (and individuals) with different, and occasionally opposing, objectives, especially on contentious issues.

It may seem counter-intuitive that distrust and conflict are what brings together collective governance, but when the parties realize that they have something to gain, and a lot to lose from not collaborating, it becomes essential. It is the tension between these different stakeholders that brings them to the table. It is the continuing tension that keeps them at the table. If the discussions are not difficult, the group is not discussing the right issues.

Each stakeholder must have much to gain and much to lose. Even then it often takes brave and innovative leadership (see Chapter 6), or a crisis, to accept the need to change an existing way of operating. Many leaders refuse to sit across the table from those actors that they consider are out to challenge, oppose and undermine them.

The companies might be considered to be the group with the less obvious compulsion to engage directly with citizenry. There are not many cases in which companies will see a reason to come to the table. This might not be out of an active desire to avoid engagement, but merely that they see their role as "doing business" not discussing public policy. Therefore, the business case for collective governance needs to be strong. The issue must be complex or urgent, fuelled by anger, conflict and campaigning, and the companies must have a lot individually and collectively riding on resolution. For example, collective action is unlikely to succeed if the sector has too many disparate companies for any single or small group of companies to be under significant reputational pressure.

Massimo Mantovani[2] cites five advantages of collective action for business:

1. Maintaining a level playing field in an increasingly global market

2. Improving public trust in business by increasing transparency and accountability

3. Ensuring that business acts in accordance with the legal framework and even contributing to change in the regulatory system

4. Improving efficiency and avoiding overcharging

5. Reducing space for corruption

This makes a good list for what there is to win, but the obverse must be true: companies must have a lot to lose if they do not collaborate.

2   Mantovani, 2012.

This generally is only the case where the barriers to entry for companies are high. In the offshore oil sector, for example, it is not easy to set up a rig and start drilling. It takes huge sums of money, exploration, technical skills and equipment, lots of bidding and contract negotiation, massive logistics, and engagement with government. Onshore might be harder because it involves engagement with the community. Michael Warner and others have written extensively on the cost–benefits to business of good community engagement well beyond the reputational (Warner and Sullivan, 2004).

This means that, in the extractive sector, there tend to be just a few companies, characterized by large capital expenditure and a long lead time until they know if their investment has been profitable. Furthermore, these are big international companies with shareholders looking at share prices. If any of these companies has a reputational scandal, the cost can

What got everyone round the EITI table

The EITI was formed by a perfect storm of anger arising from academic work, NGO campaigning, and pressure that companies felt ill-equipped to deal with alone. Everyone had a reason to be around the table:

- Citizens wanted a better deal, a place at the table to discuss public policy, and information.

- Governments wanted more revenue and investment, and to polish their reputation.

- Companies wanted a level playing field, a more predictable, stable environment, a social licence to operate, and a chance to demonstrate their social and economic contribution. Oil and gas, and to a lesser extent mining, had a few high-profile operators in each country with long-term, high-capital, high-risk operations—they could not afford for anything to go wrong.

Yet, even with these reasons, the chance that strong-willed campaign organizations such as Global Witness and the then Revenue Watch Institute would sit around the table with hard-nosed oil companies, while continuing their campaigning, seemed remote. No wonder the *Economist* described the EITI back in 2008 as a "curious coalition".

The enduring conflict between the parties, together with the continuing mismanagement of the extractives sector in many countries, continues to give everyone a reason to sit around the table. The campaigning by NGOs has done more to build company support for the process than any direct recruitment efforts by the EITI International Secretariat. Similarly, the companies are to be applauded for engaging so openly and constructively while receiving sharp public criticism.

be disproportionate to the damage (as was arguably the case with BP in the Gulf of Mexico oil spill). In short, if anything goes wrong, they have a lot to lose.

The authors have seen many attempts towards collective governance, in which the business case for companies to be around the table was too weak—not enough anger, not enough urgency, not enough potential for reputational damage, and in a sector which is simply too low cost and too dispersed.

There is, of course, a third precondition: cost-effectiveness. Though the authors have nothing new to say on this matter, it should be noted that, in matters of governance, the opportunity costs or counterfactuals are so difficult to define in a simple cost–benefit analysis. How do you cost a conflict or a policy reform or a misinformed debate that didn't happen?

Lessons from the chapter

The collective governance route is likely to work best when the following conditions exist:

- Clear government failure where there is a lack of trust in both government and companies.

- A problem that requires long-term co-operation, not a one-off deal.

- All players have something to gain and a lot to lose.

- Enduring conflict to continue to get the players around the table and to keep them there.

- A manageable number of main companies.

In short, collective governance is an expression of crisis.

4
Build trust *through* building momentum: just get on with it

The first part of this chapter's title—building trust *through* momentum—is not a straightforward concept. As noted in the previous chapter, collective governance is built on the precarious foundation of distrust. Collective governance usually comes on the back of zero or low trust and a lot of animosity between the actors. Indeed, it is the authors' experience that disagreement among actors is an essential starting point and a driver throughout the process. Those involved are representing often opposing interests. Most have been chosen for their passion and often have a strong capability to harness arguments to oppose and wrong foot others. There might well be a majority of actors who want the project to fail or cannot see how they can compromise on these issues. Even if it is not unusual in a political context to have governing bodies with different objectives represented, it is unusual to have so many different objectives and strength of opinion around the one table as is found with collective governance. This means that getting started is particularly challenging and progress is incremental and often slow.

At first, the main reason to have collective governance is to stop people with different interests shouting at each other: i.e. convening in order to build trust. Most of the academic literature on collective governance advises that time is taken first to simply establish trust. The authors do not subscribe to this view. Trust without substance is fragile and ultimately unsustainable. Getting the balance between building trust and "getting on with it" is one of the hardest challenges for a governance entrepreneur. The authors' advice is not to wait until trust is established before undertaking substance—**where progress on substance is made, trust will follow**. This has some important and radical implications. Because getting started is so difficult and because time can be wasted on ephemeral matters, this chapter will begin on three points to facilitate progress: agree on general rules of the game—consensus, but not necessarily voting; don't get hung up on the legislative process and whether it should be a voluntary or mandatory initiative; and perhaps most importantly, accept power imbalances.

EITI and the power of convening

In the early days of the EITI, these multi-stakeholder platforms for dialogue and oversight were, in themselves, the chief success of the EITI. There was strong evidence that bringing hostile parties to the table in itself reduced tensions, particularly in post-conflict environments such as Liberia, the Democratic Republic of Congo, Afghanistan and Iraq. There was great commitment to implement and a lot of countries with EITI multi-stakeholder groups were giving an almost unique platform and voice to a variety of stakeholders. In some countries the EITI process even protected activists (see p. 57). However, by the end of 2008, there were less than a dozen actual EITI reports, and most of them were poor quality. In other words, groups were meeting but lacked the data to inform their discussions to build consensus. There was a temporary "trust", but it was a precarious, unsustainable position.

4.1 Consensus not voting

To make progress among varied interests requires a working understanding of the concept of consensus. The authors' experience is that it is best to avoid over-formalizing this concept by wasting time on laborious and hypothetical voting rules which can drain away trust.

There are many definitions of consensus. Often discussions on how to make decisions get drawn into protracted debates on voting arrangements. While it might be useful to have this conversation, especially where trust is low and power relations unequal, it is critical that the conversation does not get stuck. The bottom line about voting—whether or not it is allowed for in the regulations—is that it should never happen. If a vote is needed, consensus has already broken down and trust has been lost.

The word "consensus" does not mean the same as everyone agreeing. For consensus to be attained, not all parties have to agree, only that all parties accept the way forward. This distinction becomes important often to

protect individuals who have to toe a line from their organization or constituency, sometimes even for legal reasons, but also recognizes that theirs
is a minority view that could hamper progress. It can be useful, for example,
for minutes of meetings to record individual dissension from the "agreement", even if it seems contradictory to have a decision by consensus with
records of dissent.

There is a science to consensus building. It is time-consuming but it is
core to the success of collective governance. To avoid a lack of consensus,
a significant percentage of a governance entrepreneur's time will be spent
on one-to-one conversations with the members of the collective governance group and their constituencies, providing background information,
facilitating working groups and committees, getting opposing views to talk
to one another before coming into the board room, hosting networking
opportunities to relax board members with one another socially, and so on.
This is discussed in more detail in Chapter 7.

The EITI and voting

The EITI International Board has rules for voting[1] but, in over 30 meetings,
has never had a vote. There have been occasions when a vote has been
discussed and then the board has adjourned for smaller group caucuses to
seek resolution. On one occasion, the board took a "mock vote" in which
members unofficially voted to see where a voting process would leave them in
order to see if it helped forge a clear way forward.

Yet every one of these meetings has involved hundreds of prior phone
calls, meetings, draft papers and social events. The board meetings
themselves never happen in isolation. At the very least they will follow a day
of constituency and committee meetings, and a reception in the evening at
which members with differing board views will be informally huddled together
to debate their points in a social setting. A lot of trust and progress can be
garnered from investing in this "social capital". It is rarely in anyone's interest
for major differences to come to the board table, but nor is compromise easily
attained.

The real test of a governance entrepreneur is their ability to concoct
ingenious, acceptable and balanced third options which do not compromise
the credibility of the process, and their facilitation and mediation skills to get
these options accepted. This is innovative diplomacy at its trickiest.

1   Article 15.5 of the EITI Articles of Association (EITI, 2013c): "If a vote is required,
    resolutions are adopted by a qualified majority requiring 13 votes to be cast in
    favour of the resolution, and must include the support of at least one third of
    the votes of the Board Members from each Constituency."

4.2 Don't get hung up on the legislative process and whether it should be a voluntary or mandatory initiative

Conversations about whether the process needs legislative underpinning are inevitable. For example, should there be stronger rules against corruption, better laws protecting human rights, or legislation to require more transparency and reporting? In addition, it is not unusual that many stakeholders will push for all the collective governance processes and institutions to be anchored in law before convening. Countries with a heritage of Napoleonic code tend to be more law-based than those with an English common law background. Countries with a strong American background tend to be more litigious. These heritages might affect the strength and nature of the calls for legislation, but the demands will surely exist everywhere. These demands and conversations can be useful, but should not be all-consuming.

Furthermore, it is embedded in the DNA of campaigning civil society organizations to achieve tangible "wins", such as passing new legislation, making new requirements, etc. Similarly, it is embedded in company DNA to resist these unless it can be clearly demonstrated to reap competitive benefit—light regulation allows them to be more nimble and creative. The history of corporate citizenship has been built on this tension.

In the view of the authors, protracted debate about legislation can be a distraction that drains energy, time and trust. The key issue is to get started and, for that, the process simply must not have significant legal constraints or other obstacles. For example, in the case of the EITI, in most countries the process only required a relatively straightforward ministerial decree requiring all government agencies and companies to comply with the EITI reporting requirements, and overriding any existing secrecy provisions or confidentiality clauses. In other cases, a simple memorandum of understanding (MoU) between all entities has proven sufficient, at least to begin with. There has never been a case of a company or government agency citing legal reasons for not complying with the EITI, yet there have been lots of cases of non-reporting entities being found in breach of ministerial decrees or MoUs. As Kofi Annan says, "Building trust is harder than changing policies—yet it is the ultimate condition for successful policy reform … Mutually beneficial agreements are the only ones that will stand the test of time."[2]

2   Annan, 2013: 7.

This is not to suggest that the authors oppose legal underpinning, just that it need not be a precondition for starting a process. Often, the process can be a stepping stone to drawing up better laws or reinforcing existing or new legislation. For example, experience shows that disclosure legislation written after a few years of EITI implementation has proven more robust than that written at the beginning of a process. Legislation written at the beginning of a process tends to be long on descriptions of the mandate, role and composition of the collective governance institutions, but short on what they actually exist for. It can be even stronger if it is integrated into wider legislation, like a mining law, rather than a stand-alone EITI Act. On the other hand, the authors have observed that the drafting and enacting of an EITI law is important for raising the profile of the process, as it contributes towards ministerial and wider political ownership. Such ownership can in turn contribute towards improved links with wider reforms.

While legislation might not be necessary at the beginning of the process, keeping abreast and co-ordinating well with legislation in the same field is critical for taking full advantage of complementarity and reinforcement.

If there is no law underpinning the collective governance process, then critics and sceptics might make accusations that it is voluntary and toothless. But this accusation is somewhat confused, since collective governance processes are usually neither voluntary nor mandatory.

Collective governance derives from government failure—an inability to pass or enforce laws. It relates to a lack of trust or information between government, companies and civil society. Therefore a simple legal or mandatory solution would not resolve the issue. Though legislation might be part of the solution, citizens demand a place at the table to monitor the process and discuss implications. Whether or not this process somehow gets embedded in law, the trust-building focus of collective governance actions goes beyond the simple mandatory/voluntary battle lines.

The EITI: beyond mandatory/voluntary battle lines

The EITI is often described as a voluntary initiative. This description has been common among NGOs campaigning for listings requirements, and among some companies wishing to emphasize that the EITI is not legally binding on them. It is confusing to define the EITI as voluntary and often stems from a failure to define for whom the EITI is voluntary. The confusion comes down to the misunderstanding by some that the EITI is a corporate social responsibility standard. While companies' support and consideration of the EITI are often part of their corporate responsibility, it is governments not companies that implement the EITI.

Furthermore, of course, it is voluntary for governments to implement the EITI. Governments are sovereign and exercise the right to engage or not with any process unless they are bound through treaty or otherwise under public international law. In almost all implementing countries, the commitment to implement the EITI has been enacted or decreed in some way. In Nigeria and Liberia, for example, there are specific EITI laws. Future governments in those countries would have to repeal these laws should they wish to stop implementation. In other countries, such as Norway, Ghana and Sierra Leone, the minerals and mining or petroleum laws include specific clauses on EITI implementation. In other countries, there have been Presidential or Ministerial decrees and, in others, memorandums of understanding. What is clear is that in every case there has been strong direction by the government for agencies and companies to comply with the requirements of the EITI. And if this has not been effective, a country cannot be deemed EITI compliant.

Therefore, a company cannot decide whether or not to report according to the EITI in implementing countries. It is irrelevant whether a company operating in Timor Leste or the Democratic Republic of Congo has expressed its support for the EITI, or whether it is state owned, British, French, Malaysian or Chinese. If it is operating in an an EITI implementing country, it will be required to report.

Furthermore, in some jurisdictions, including the United States and Europe, there are requirements for companies about disclosure of payments. While these are often referred to as mandatory processes, they too, of course, have a voluntary dimension, as companies can decide whether and where they wish to be listed or registered. These laws and the EITI are mutually reinforcing, but it is only the EITI that ensures that there is a forum to debate the data in the host country. Keeping revenue data in Washington, Brussels or Frankfurt does not necessarily improve the public debate and accountability in Lima, Jakarta or Kinshasa.

4.3 Accept power imbalances

While it is desirable to get the right people around the table and to have a balance of interests represented at the beginning of any collective govern-ance effort, it is more important that the process does not get bogged down in endless discussions about representation, profile, mandate, and so on. Power balances are dynamic in nature, most especially through a collective governance process. It is a process and forum for addressing imbalances, building capacities and empowering as you go along. Where the process ends up is more important than where it starts. If you are discussing the right things, the right people will come to the table.

It is inevitable that relationships in collective governance groups are unbalanced. This leads to a complex power play in which consensus building is a constant challenge. There will be debates about, for example, representation between constituencies; regions; North and South (for global initiatives); implementers and supporters; gender; age; background; are they representing individuals, their organizations, or their wider constituency;[3] different ways of working; voting rules (see above); funding; training opportunities; etc. The patience with discussions about representation and other organizational matters also varies with the different constituencies: private-sector representatives are less likely to be willing to spend time on procedural deliberations than those representing governments and civil society.

Some discussions at the beginning are important for building trust, but too much will lead to the initial enthusiasm for collective governance giving way to stagnant process discussions, and the group is likely to collapse with the first turnover of participants. A collective governance effort should be careful not to get bogged down organizing itself, and forgetting to get on with the job for which it was established.

This is particularly challenging as the weakest participants in the group, at least at first, will tend to be far more comfortable talking about governance arrangements than they will about substance. It might take several years before the discussions around the table move on from presentation of papers to their substance, or from how to represent different constituencies to what to do about the issues. Eddie recalls going to a civil society meeting about the EITI in Nigeria in 2008, in which the whole meeting was taken up discussing what representation to give to disabled groups in the multi-stakeholder group, rather than discussing the more than US$5 billion identified by the Nigeria EITI report as owed to the Nigerian government. Without wishing to dismiss the importance of disability rights, that civil society group, generally uncomfortable on technical issues about the oil sector, was failing to act on the shocking facts in the report. This shows how opportunities can be lost if representation issues are allowed to have too high a prominence.

---

3   As an example, on the EITI Board, the representatives of countries, civil society organizations and companies have a duty to "act in the best interest of the EITI … at all times" under Article 13 of the EITI Articles of Association (EITI, 2013c). Nonetheless, the board members remain the representatives of their respective institutions, each of which has its specific interests.

Conversely, when there are disagreements over substance and outcomes, there is often a tendency by those who feel threatened to take refuge in faulting the process and procedures. While there should, however, never be an excuse for poor process, the focus on substantive outcomes should not be lost. In many countries, EITI reports have highlighted unexplained payments, unjustified allocations of contracts and clear non-payment of taxes. In most of these cases, there has immediately been a backlash of questions about governance and secretariat procedures. Auditors and financial assessors are called for, accusations made, fingers pointed. In the cloud of smoke put up by wrongdoers, it is necessary for governance entrepreneurs to keep all eyes fixed on the chief misdemeanour, even while fingers might be pointed in their direction.

Disagreements on process and on substance can be destabilizing. Michael Jarvis of the World Bank Institute asks:[4]

> The enthusiasm and likely confusion of the initiation phase gives way to concentration on an institutionalized process through the implementation phase, but that can only go so far before questions of effectiveness arise, before there are calls to adjust for changing circumstances, and a range of potential destabilizing factors kick in such as turnover of players involved (individuals and organizations). Does this mark the equivalent of [a collective governance] mid-life crisis?

4.3.1 Power relations are dynamic and collective governance is part of the changing relationship

The bad news for the governance entrepreneur is that power will never be in equilibrium. The process needs constant management and cajoling. The complexity of the challenge never lessens because the balance of interests is dynamic. Someone will always be the weakest, and someone else the strongest. While civil society power might be on the rise, it is very different from country to country, it is not coherent and it is changing—indeed, and crucially, it is collective governance itself that is contributing to the change.

Civil society is there as a representative of what might be a democratically limited wider society. Consequently they are often likely to be lacking power, inexperienced and not knowledgeable about the sector. Strong members of civil society might well be drawn to other topics. Collective governance, much like other collaborative efforts, often makes unreasonable demands on the limited capacity and capability of civil society. Even in

4   Jarvis, 2013.

collective governance processes, civil society can easily be co-opted or marginalized by the government, or simply lack the capability to hold government and companies to account.[5] This has certainly been true in some EITI cases. Opoku-Mensah concludes from a CIVICUS survey of African civil society that "the weakest area of impact is at the level of holding the state and the private sector to account".[6] This is not just the case in sub-Saharan Africa, but also in other countries in which political space for civil society has been limited. In a generally supportive article, Virginia Haufler states that EITI processes can at times be a "triumph of form over results, with real power remaining in the hands of government and corporate elites".[7] However, as with protection of civil society (see p. 57), collective governance is a key mechanism for improving the capabilities of civil society.

The Publish What You Pay coalition and the Natural Resource Governance Institute have played a critical role in contributing to the strengthening of local civil society to act as a watchdog for the extractive sector. They have done this across many countries. This has increased the capacity of ordinary citizens, parliaments and media outlets to provide effective oversight in the area of natural resource extraction many times over. It would simply not be correct to say that civil society, once co-opted or marginalized, cannot eventually hold governments and companies to account through the EITI process. But it will always take skilled co-ordination to adapt to the uneven and changing power balances within any multi-stakeholder forum.

For collective governance to work, capacity building is required for all stakeholder groups, not just civil society. No group is likely to be used to this way of working. This is particularly true for governments for whom the process is ultimately meant to strengthen their institutions. They need support for how best to use the process—in the case of the EITI, to increase investment and revenue, build a strong extractive sector and ensure that the benefits are more equitably shared. Facilitating capacity building and mutual learning is a critical part of a governance entrepreneur's job. It might include anything from staff exchanges and peer learning to more formal training and guidance materials, or board and other stakeholder missions to implementing countries.

A governance entrepreneur needs to be constantly alert to how to persuade important stakeholders that this collective governance work is more

5   Carbonnier *et al.*, 2011.
6   Opoku-Mensah, 2007.
7   Haufler, 2010.

important, engaging and exciting than their other priorities. Global networks, training and opportunities to travel are, for most stakeholders, rightly or wrongly, an important part of this assessment. The governance entrepreneur will therefore need to make shrewd judgements not just about how to dispense and provide such capacity-building opportunities, but also about how such opportunities are viewed as incentives. In many countries, the opportunity to go for training in say, Germany (much of the EITI training is provided by the German co-operation agency, GIZ, some in Germany itself)

The EITI Board and implementing countries

The EITI is implemented by countries. Yet this is the constituency on the board with the weakest voice. Before each board meeting, civil society has at least one day of meetings and comes prepared and usually aligned on all topics. Oil and gas companies have regular phone calls, as do the mining companies and the investors. The supporting countries have phone calls before the meetings and then a face-to-face meeting the night before to confirm lines.

It is different for implementing countries. They do not necessarily have naturally common positions—there is no reason why Azerbaijan and Timor Leste would necessarily take the same view on matters. Why should a compliant country support a non-compliant country seeking an adaption from the standard? Why should a candidate country support another to seek candidature? Yet this lack of coherence makes the most important group also the weakest group. This is not healthy for the long-term interests of the EITI. The International Secretariat has worked hard to address this by:

- Facilitating annual meetings of all implementing countries with the objective of updating them, providing space for their debate and best practice, and building up a network of mutual interests.

- Issuing regular circulars to the implementing countries on relevant upcoming issues and news.

- Facilitating phone conferences before board meetings.

- Facilitating the process to elect the representatives of implementing countries on the EITI Board. This involves breaking the countries down to more coherent sub-constituencies such as a Francophone African group (though that brings other challenges of creating bloc interests and demands for special treatment).

- Hosting regional training events.

Yet this remains a chief challenge for the International Secretariat, and the interests of one implementing country can often work against the interests of another.

is very attractive. While visiting Germany should not be the end in itself, it also should not be ignored as a way of engaging important stakeholders.

4.3.2 Rules vs. reality: encouraging progress while discouraging organizations gaming the system

Often non-reforming countries or jurisdictions are where collective governance processes are most needed. It might be the only platform that exists for civil society. It might give space and profile for reforming instincts in government to catch hold. The strength of collective governance is that, by definition, it is not just implemented by government: it is also implemented by civil society organizations and companies. Often where the regime is repressive, a collective governance process is the only platform available for dialogue or reform efforts among other actors.

For a global standard such as the EITI or the High-Indebted Poor Countries (HIPC) process for debt write-off, the difference in entry point and compliance is a powerful tool. This locks in the stakeholders and encourages them to continue to make progress. It is sometimes pleasantly shocking to see how often countries unwittingly commit to a standard before fully understanding the implications for their behavioural change. This is often skilfully managed by reformist governance entrepreneurs inside a regime.

However, in accommodating wider demands and challenges, the collective governance process needs to be conscious of its credibility—"the value of its currency". Even with robust rules and processes, there is a danger of being used to present a whitewashed façade hiding a tarnished reality.

It is essential to acknowledge that collective governance cannot magically create political will for reform in governments where there is none. Assessing real political will is both complex and a perennial challenge for development initiatives. Foreign and development policy has never settled on an agreed balance between encouraging and keeping difficult governments inside the tent, and throwing them out. The situation is made more complex by the fact that governments are made up of individuals, often of vastly different attitudes.

Ultimately, the authors cannot present from their experience any magical answer to how high to set the bar for keeping countries (or others) in or out of the process, nor how to set the rules to safeguard against exploitation. It is a matter of vigilance and judgement. Even with collective governance, that remains the most vexed question in development. Robustness of rules might be one thing, but the literal letter of the rules needs to be balanced with good judgement about their intent. The EITI faced this judgement when the credibility of its rules was on the line in 2010.

The EITI Berlin board meeting

In May 2010, the EITI Board met in Berlin. It was to be a seminal moment in EITI history. The board had agreed two years previously that all candidate countries would have two years in which to undertake a validation or be delisted, unless they could cite "exceptional and unforeseen circumstances". If the validation assessed that "no meaningful progress" (i.e. no EITI report) had been achieved, the country would also be delisted. The deadline for the first 21 countries was at the Berlin board meeting. Only three of the 21 countries had met the deadline, of which Azerbaijan and Liberia had already been found compliant, and Mongolia had made meaningful progress. The fate of 18 EITI countries was in the balance: 17 had requested an extension citing exceptional and unforeseen circumstances, and one had sought a voluntary suspension.

To the average observer it seemed unlikely that so many countries could have *exceptional and unforeseen* circumstances. Either they had to be delisted, leaving the process bare, or the EITI was not true to its own rules. The credibility of the process was on the line.

Yet from the inside, there was an encouraging story—deadlines were working. Eighteen of the 21 countries had disclosed revenues from oil, gas and mining production, 16 of them in the previous nine months. Eleven of these had produced these reports for the first time. Validation had begun in all but four of these countries. Implementation had been a learning process for all. Validation, in particular, had proven to be a complex and time-consuming exercise, but was also successfully highlighting the strengths and weaknesses in EITI implementation. Given all this progress, the board had to decide how strictly to impose their rule, mindful that the world was watching. It was critically split.

After much painful discussion, the board agreed to grant extensions to 16 countries. It delisted Equatorial Guinea and Sao Tome and Principe. Some international NGOs issued statements criticizing the board for its weakness. Others praised it for pragmatism, given that the rules had been written before experience had shown what would and would not be reasonable. Either way, the EITI had demonstrated that it had teeth by delisting non-performing countries, something that other international standards bodies, such as the Kimberly Process, had been unable to do. Four years later, all of those 16 countries had become compliant, except Gabon which was subsequently delisted in 2013.

In a twist to the story, it was the week that volcanic ash fell on Europe and halted all flights. Most of the board members who had been locked in heated and emotional arguments over the week were stuck in Berlin with one another for an unknown further period. It was clearly an exceptional and unforeseen incident. Previously conflicting groups quickly organized around logistics. There was a particularly pleasing story of the representatives from the oil company Total, the mining giant Areva, the NGO Secours Catholique and the French government, all renting a car back to Paris: collective governance literally on a journey together. Heated exchanges were mixed with bonhomie and spirit of adventure.

The EITI and falling below the standard

An EITI compliant country cannot simply stop or diminish its process and hope to maintain its compliance. Not only is it subject to validation every three years, but also anyone can raise their concerns to the board at any time about the country falling below the standard. If the board agrees that the implementation has indeed fallen below the standard, it can ask the country to undergo a new validation or secretariat review.[8] On the basis of all this, the board can designate a previously compliant country as a candidate, or suspend or delist it. Azerbaijan became the first country to undertake validation under the standard in 2015 and, as a result of the validation, became the first country to be downgraded.

4.3.3 The good of small wins

Politics is the art of the possible. Curious coalitions, lack of trust, uneven balances of power, narrow consensus, no long-term objectives and a need to make substantive progress can all make for an often unsatisfactory cocktail of compromise. One of the costs of keeping an unbalanced set of powers around the table is that they all expect a regular diet of "wins" from the process, whether or not their case is strong. This can be boosting for the individuals themselves or can be something that they can take back to their organizations and superiors to show the value of the process. There will be times when the governance entrepreneur, as with the other stakeholders, will have to accept suboptimal, or what the authors simply call "stupid", decisions in the interests of making progress. Furthermore, the governance entrepreneur will have to make the case for some of the stupid ideas themselves as part of a mediation role, i.e. you accept X from stakeholder A, and

8   A secretariat review is a lighter touch version of the validation, in which the International Secretariat provides to the board an assessment of specific concerns or corrective actions, rather than an assessment of all the requirements.

I will try to deliver you Y from stakeholder B. Some of these stupid ideas can be remedied later, some will become irrelevant and some will continue to haunt. As in politics, there is sadly no magic formula for knowing which one is which, but keeping a focus on maintaining progress and balance between stakeholders often means giving small wins to stupid ideas.

The EITI and the art of small wins

As in all political processes, the EITI International Secretariat often knowingly had to take a "stupid" decision as part of the compromise needed for progress, hoping that they could remedy them later. For example, there were many unsatisfactory elements of the 2009 and 2011 versions of the EITI Rules. For a long time, the board kept a strict interpretation of the concept of a "maximum candidature period". This meant that if a country had not progressed to become compliant after a certain period of time, it *had* to be delisted. There was good logic behind this, in that it prevented countries permanently squatting in "candidatureland", i.e. riding on their reputation of implementing the EITI without making any meaningful effort to achieve compliance. The EITI deadlines on when to do validations were, and continue to be, a highly effective way to ensure progress. Validation reports often appear at five minutes to midnight on the deadline date. Almost half of the EITI reports in 2014 were published in December prior to the end-of-year deadline for 2012 data. Many board members were insistent that this maximum candidacy period rule should be stated clearly, firmly and rigidly applied. However, rigidly applied the rule "locked in" board action and therefore distorted their judgements. For example, if a politically and technically complex country, such as the Democratic Republic of Congo, was approaching its deadline having finally pieced together an enthused, functioning multi-stakeholder process and produced startlingly important and new information on the sector, but failed the standard on one minor technical point, the country would face the humiliation of delisting. As another example, if one company out of 40 had not responded, or one government agency had not got its figures attested by the national audit body even though it matched with the company figures, it would lead to delisting. The chances of getting the country back into the process could be remote. Clearly the rigid "lock-in" was unwise. Many members of the board and the International Secretariat knew this, but knew that this had to be conceded in order to get broad support for other reforms in the EITI Rules. When country cases relating to the rule were brought to the board, it led to tortured debate, distorted arguments and eventual fudging. The rule was soon afterwards revised—the deadline remains, but delisting is not a *necessary* consequence of failing to meet it.

The EITI and civil society

No issue in the history of the EITI has created more debate than its role in relation to civil society. Where there is corruption and poor governance, there are often human rights abuses. Children may not be able to exercise their right to education, poor parents might not afford the fees demanded by teachers, or campaigners against corruption may have their freedom of speech curtailed. The remedies against corruption cannot be considered in isolation from human rights. There has to be some kind of human rights enabling environment for those campaigning for transparency and fighting corruption. The EITI has been criticized by many for not doing more to explicitly promote human rights in countries such as Azerbaijan or Ethiopia. There are many important and complex links between the promotion of accountability and the respect for human rights. Transparency alone does not improve governance—some basic human rights have to be respected for collective governance efforts like the EITI to be implemented in a meaningful way. The EITI cannot and does not ignore the broader human rights agenda, but that is not to say that transparency will only lead to change in "free countries". The EITI demonstrates that transparency can still contribute to change in countries with poor or worsening human rights records.

It is a challenge faced by many other collective governance efforts. The Open Government Partnership is grappling with how to respond to perceptions of closing civil society space in some member countries such as Hungary and Montenegro. The Kimberly Process was almost felled by failing to take action against Zimbabwe over its human rights record in diamond mining. It is a complex topic on many levels, as an active, independent and engaged civil society is required for an effective process, yet these concepts are not absolute and the process itself can improve and protect the lot of civil society.

Here the EITI is assessed as a protector and as a capacity builder. Then this box looks at the conclusions for its policy on human rights with reference to a couple of case studies. Finally, it is important to explore the role of the global coalition for civil society in this field: Publish What You Pay.

Collective governance as a trust builder: the EITI and protection of civil society

In many countries, the very existence of an EITI multi-stakeholder group can give reformers, activists and enlightened companies a platform in challenging environments long before any data is produced. Anthony Richter, chair of the governing board of the Natural Resource Governance Institute and former EITI Board member, said: "When the [Rapid Response] Committee [of the EITI Board] intervenes it does so with considerable force [and] interventions have helped, if not to correct general patterns of human rights abuse, then to stop harassment in specific cases, using the influence and authority of the Initiative".[9] There are many such examples of the EITI being cited as having protected civil society activists when they have faced persecution for campaigning on extractive industry governance.[10]

The EITI and improved capacity of civil society throughout the process

A country's first EITI report is rarely a document of great substantive importance. It tends to contain discrepancies between company payments and government receipts—usually due to misunderstandings about categories, accruals and cash, payment dates, currencies, etc. The reports are often set out poorly and difficult to read—they lack context or narrative. Sometimes the reports fail to explain the complex payment processes. A weak and ill-informed civil society finds it hard to engage on the scope, detail or analysis. It might feel marginalized from the process and even feel that it is being unwittingly co-opted to endorse murky and cosy arrangements between its government and companies. For their part, some governments have signed up to the EITI without any real commitment to reform, thinking that they are in for an easy ride because civil society cannot understand these issues.

In the second reporting cycle, however, civil society tends to demand a clearer scope and context. An unreforming government finds it difficult to suppress this request through "conventional" means (see above). Whether civil society gets all that it wants or not, the second report then tends to be of far higher quality—in general, countries tend to pass the EITI Standard on their second or third report. The context and data are clearer and civil society begins to know the right questions to ask.

The third reports tend to be of even better quality and with deeper information, better set out with more context and more analysis. Linking well with their international networks, peers in other countries, media, academics and the international community, civil society representatives ask tough, incisive questions and hold the institutions more to account. Formal training may have been an element of this development, but it is more likely that three years of slow engagement with the sector and with the process have built capacity to understand and appreciate the challenges of governing oil, gas and minerals.

In addition, of course, by focusing on improved readability and usefulness of all EITI reports, the EITI Standard (launched in 2013) should improve civil society's ability to engage in the process.

9   Richter, 2010.
10   In Gabon (Natural Resource Governance Institute, 2006), Congo-Brazzaville (Natural Resource Governance Institute, 2008) and Niger (Natural Resource Governance Institute, 2009).

The implications for EITI policy on human rights

The EITI has relatively low minimum requirements for civil society to participate freely and fairly in the process for two main reasons. First, the EITI was in part designed to create a platform for change in countries where human rights standards are often below the desirable. As shown above, the process itself can often improve the voice and capacity of civil society, even in these challenging countries. It is fair to say that the establishment of the EITI process is almost universally supported by that country's civil society—it often provides a critical safe platform that might not exist elsewhere. If the situation is such that civil society is not able to participate, the EITI has clear rules and mechanisms for reviewing whether the country can continue to implement the standard.[11]

It is, of course, beyond the expertise of the authors to comment on whether some kind of minimum bar for civil society is in fact necessary for sustainable development take-off, although that appears to be increasingly questioned in some development literature. For example, Brian Levy—who is a strong advocate of the incremental pragmatic approach to governance—argues: "perhaps a seeming excess of order or a seeming excess of chaos may be less a signal that a country is off-track than part of the (medium-term) nature of things".[12]

Second, it is not easy for something like the EITI to consistently apply minimum standards for active, full, independent and free participation. Of course, some countries have prohibitive laws and regulations, but allow relatively free participation in practice. In some other countries, civil society has learned what the government can accept and there is a high degree of caution and self-censorship by civil society groups.

The 2013 EITI Standard seeks to ensure that the transparency delivered through the EITI does a better job in leading to accountability. It empowers civil society to put a much wider range of governance challenges on the table. Civil society organizations receive and will continue to need support to realize this potential.

In 2009, **Ethiopia's** application to implement the EITI brought this debate directly to the board table. Much of the international NGO community was outraged by the then recent draconian proclamation which limited foreign funding to Ethiopian organizations in the field of governance. A minority of the board argued that Ethiopia could not establish the full, free, active and independent participation of civil society in the process while this law was in place. Crucially this was supported by at least one implementing country representative. Thus consensus could not be established. In the absence of a consensus, the status quo stands—in this case that Ethiopia's application was not passed. There have since been three board missions to Ethiopia to

11 EITI, 2015a.
12 Levy, 2014: 44.

assess the situation, each citing the support of Ethiopian civil society to the application. Ethiopia continued to implement an EITI-like process outside the formal international EITI structures until it was finally accepted as a candidate country in March 2014. Some argue that countries with lower standards for civil society freedom of activity than Ethiopia never faced the same rigorous scrutiny in their applications to the EITI.

In October 2014, the board considered the case of **Azerbaijan** where, despite having become compliant in 2009, the space for civil society active in the sector was perceived by many to have closed: the bank accounts of many civil society activists in the sector were reported to have been frozen, meetings disrupted and organizations required to re-register. It was hotly debated as to whether Azerbaijan was in significant breach of the EITI Standard. The board concluded that a revalidation was required to test exactly this question and whether it had slipped on any other points. This validation was undertaken in the first quarter of 2015 and concluded that the process did not met the standard on several points, and Azerbaijan was downgraded to candidate status.

The EITI and Publish What You Pay

The EITI's future is linked with Publish What You Pay (PWYP) continuing to be an effective coalition to improve accountability at the global level and in each country. Although the PWYP is, and should be, a light touch network, its members are the chief users of the transparency brought by the EITI, carry the main burden to turn it into a tool for accountability, and are the predominant body on which national EITIs most rely for shining the torch towards their future direction. At its best the global coalition creates a network of knowledge and expertise, sharing of best practice, guidance, training and support for civil society in implementing countries. The longer a national coalition is engaged in its country's process, the better able its members are to demand and analyse the right information and to hold its government and companies to account. The closer the national coalition is connected with the global coalition, and the stronger that global coalition is, the faster its learning curve and the stronger the EITI.

However, on matters of global policy such as Ethiopia and Azerbaijan, civil society is often split. International actors are predominantly concerned that the EITI Standard is not lowered, used or whitewashed by what they consider oppressive governments, and the national actors often cling to the process as their best platform for expression, protection, capacity building and support.

For organizations such as PWYP, this brings almost unbearable countervailing pressures. While mandated to bring together the interests and voice of citizens most affected by natural resource governance, they are funded largely by rich northern-based NGOs with different interests. To navigate this terrain requires that PWYP has strong governance mechanisms which give sufficient voice to the affected citizens and independence from the funders. Progress on independence has been made recently with PWYP no longer "hosted" by the Open Society Foundation, but the tension is likely to continue. It might come to a point where PWYP will have to ask itself if it can be both a coalition as well as an advocate for such varied interests.

The role of companies in the EITI

While national governments are ultimately responsible for implementing the EITI, the co-operation and support of companies has been important in maintaining momentum and defining scope. In many countries, it has been the companies that have pushed the government to implement. At the international level, they have been critical in defining the mandate of the EITI, including many of them arguing for the transparency of licence processes and registers, beneficial ownership, and subnational and social payments. Almost 100 major oil, gas and mining companies are "supporting" the EITI at the international level by expressing their support of the EITI Principles on their website and, in most cases, paying a voluntary contribution towards the international management of the process. A further almost 100 institutional investors support the process.

Drawing on the work of Martin Tisne, Director of Policy at Omidyar Network, and others, the authors identify six different roles for extractive companies in the EITI, which might be relevant for other collective governance initiatives:

1. **Transparency advocates**: particularly by promoting the implementation of the EITI with the host government and/or informing their investor decisions based on assessments of licence to operate and political risk. Companies often quietly approach their host government to consider the EITI and corporate bodies, such as Chambers of Mines, often drive the process.

2. **Shapers of the EITI data**: by influencing the scope of the EITI at the international and country level and making it more accessible, e.g. the EITI company board members.

3. **Beneficiaries and users of EITI data**: using the data to inform and influence their business decisions. As Aasmund Andersen has written: "Investors ... rather than stakeholders, are the primary users of the public data for their decision-making and due diligence purposes".[13]

4. **Forum for peer dialogue**: the company constituency of the EITI is broken down into oil and gas companies, mining companies and institutional investors. Each of these have regular sub-constituency meetings usually before and after each board meeting, not only to discuss EITI matters, but also to caucus on wider governance issues in the sector, such as establishing joint company positions on the proposed mandatory reporting legislation in the United States and European Union.

13   EITI, 2013b.

5. **Reporting on their payments and other activities**: as well as responding to EITI reporting requirements, they might use the company constituency forum to promote best practice and cutting-edge work, such as Tullow's project-by-project reporting in May 2014 in advance of the EU Accounting Directive requirements.

6. **Providing technical and financial assistance and advice** to governments and civil society on governance of the sector. For example, BP, Chevron, Shell and Total provided training to Burmese civil society about oil contracts in October 2014.

The new normal

However, it should not be forgotten that companies perform the above roles primarily for the private interest. This is not to suggest that the public and private interest are necessarily in contradiction. Indeed, the aim of a collective governance initiative should be to find the areas in which the interests overlap, recognizing that these areas move as the public debate moves. In the authors' experience, the chief concerns of companies in a volatile sector, such as oil, gas or mining, is that the rules be stable, predictable and apply to all companies. The EITI Rules, on the whole, provide stable and predictable rules which apply to all companies.

For example, in the early days of the EITI many companies—especially in the oil and gas sector—saw it to be in their interest for data to be available on the total aggregated revenue to the state. They argued, however, that it was against their competitive interests that it be disaggregated so a citizen could see how much each company had paid, let alone how much was paid for each project. They said that they would only do it if the state required. As more and more countries moved to publish disaggregated data, it became clear to these companies that this was "the new normal" and that it was not threatening if applied to all companies within a certain jurisdiction. It was better for them that all companies publish their payments than for it to only be applied to some. In 2011, payments broken down by company became a requirement of the EITI.

What many companies have found so challenging about the US and EU mandatory disclosure requirements—especially in respect of a requirement for payments to be broken down to the project level—is that they would only apply to those that are listed in the United States or European Union, and would not therefore apply equally to all companies operating in a country. For example, the requirement would not apply to domestic companies, private companies, state-owned companies, small and medium-sized companies, foreign companies from other jurisdictions, etc. Consequently the US and EU companies—alongside civil society—pushed for the EITI to include a requirement for project-by-project reporting, in line with the US and EU legislation.

Similarly, the major companies have pushed for requirements about beneficial ownership. Presently, the ownership of listed companies is effectively publicly available by monitoring their share ownership, but non-listed companies might well have opaque ownership. This opaque ownership could be hiding dubious overseas tax arrangements or political connections that give those companies an unfair advantage.

Should more be required from companies?

While it is somewhat clear what companies want from the process, it has not always been clear what the process can demand of the companies. Those operating in EITI implementing countries, whether they officially "support" the EITI or not, are required to publish their payments. A company that supports the EITI and that operates in a non-implementing country such as Sudan, however, does not need to report its payments. Its voluntary support of the EITI at the international level comes with no reporting obligations other than a statement of support on their website and a short annual questionnaire for the EITI website.[14] They are invited to make a voluntary financial contribution to the international management of the EITI and most of them do. Nothing more is required from the almost 100 supporting companies of the EITI.

For those people who view the EITI as some kind of corporate social responsibility process this sounds weak. Even for those who would expect a contribution from all constituencies, this statement falls pretty low against expectations. Some have proposed that the EITI demand more from its supporting companies.[15]

The requirement for supporting companies

Furthermore, at present there is no entry level for companies that wish to become a supporter. There is no mechanism to refuse a request for a company to become a supporter. Any company, however appalling their record on transparency, can become an EITI supporting company by simply uploading a statement to its website. The board is considering whether eligibility criteria for companies should be required in partial parallel to those that are required for countries. This is not a straightforward issue as it might take the EITI into areas that it was never designed to assess. However, too many dubious

14 EITI, 2015b.
15 In his keynote address at International Mining for Development Conference in Sydney in May 2013, Professor Michael Ross proposed: "EITI 2.0 should recognize that governments AND companies are equally important (in inspiring change, and transforming mineral wealth into sustainable development), and hence should be subject to similar transparency standards that are extensive, audited, validated, and trustworthy. By failing to take on this issue, EITI and many other groups working on extractive sector governance, including the World Bank, the bilateral and multilateral aid agencies—are missing a massive opportunity. The companies have the capacity to do both enormous good and enormous harm" (Ross, 2013).

supporting companies might devalue the brand and lessen the credit that truly supportive companies should receive.

These requirements might also be extended to civil society participants where there are often legitimate concerns about whose interests they might be representing. The "Who Funds You" campaign is a useful effort to require civil society organizations to disclose information on who is funding their work.

Lessons from chapter

- Successful collective governance depends on a sophisticated consensus.

- Collective governance has to look beyond voluntary vs. mandatory battle lines, and instead focus on shorter term actions leading also to trust building.

- Don't get bogged down by discussions about representation, profile and mandate. Get over it–the right people will come if the right issues are being discussed.

- On the other hand, accept some focus on process rather than substance, especially from the weak who are unfamiliar with the topics of substance, and from the strong who might be uncomfortable with the content of the substance.

- Build capacity and address power imbalance by building substance.

- Attention, network building, capacity building and facilitation are required for stakeholder groups to ensure that they can contribute appropriately.

- Allow the standard to be adaptable and attractive to all potential implementers.

- Make sure that the intent of the rules outweighs the letter of the rules, so as to encourage progress while being tough where political will is missing.

- Be conscious of giving each stakeholder some sense of a win in critical debates.

5
Move the consensus from the narrow to the meaningful

Conventional development theory uses a planning model of development: stakeholders agree overarching objectives, with a theory of change that leads the planners clearly from the activities through to the results in a linear, though often complex, route. They have a shared goal, purpose, outputs and activities, often set out in a logical framework. Milestones, risks and assumptions are set out. Intuitively it makes sense to set out clearly what you seek to accomplish. According to recent studies, the conventional wisdom for collective governance initiatives remains based on the notion that they begin by setting long-term goals with clear monitoring and evaluation targets.[1]

The authors contend that **collective governance should be managed fundamentally differently.** By definition, multi-stakeholders will have multiple objectives. While governments, companies and civil society organizations may share some overarching aims, such as a world rid of corruption, they are likely to profoundly disagree on how to achieve them. A company has a fiduciary obligation to return value to its owners. A government is accountable to its electorate. Civil society organizations represent a diverse set of interests, from narrow single-issue campaign groups to global policy advocacy. The very reason why collective governance was needed in the natural resource sector was that the parties had such wide and diverse objectives. It is a lack of agreement on objectives that has brought the parties into conflict with each other. They might be able to agree facts, even possibly

---

1   Marianne and Liese, 2014; International Civil Society Centre, 2014.

analysis, but it would be wrong to expect them to agree on answers. It is unrealistic and not fruitful to expect such varied actors to agree on long-term goals. Focusing on those disagreements is likely to scupper any progress and undermine trust.

Furthermore, in the experience of the authors, companies like "technical" challenges, whereas civil society likes "adaptive" challenges. For example, in the case of the EITI, the requirement to present a report on payments and revenues was a technical issue which a collective governance forum was required to oversee. For civil society, the establishment of a multi-stakeholder forum provided the space to address a series of adaptive issues around the governance of the sector: not just how much money was coming in, but also how much money ought to have been coming in, who was benefiting, how concessions were being allocated, how terms were being agreed and by whom, how they compared to other occasions and other countries, how communities were benefiting, etc. The report and the multi-stakeholder group were entry points for exploring these wider "adaptive" challenges though collection and use of data and through debate. Thus these two constituencies were approaching the process from different angles. This pattern can be observed also in other fields.

Collective governance vs. multilateral governance: the benefits of a technical entry point

Collective governance is hard, but so is getting a bunch of governments or political parties to agree to anything. The World Trade Organization (WTO) has undergone nine rounds—the present Doha Round has been going on since 2001—without concluding a single world trade deal. Global deals have been scuppered by petty interests and domestic politics. Progress can also often be slowed down by single-country politics. The Kimberley Process on Diamond Certification was, for example, destructively locked for many years over how to handle Zimbabwe.

Be it in the United Nations, WTO or other more formal multilateral forums, progress is often slow due to entrenched positioning by governments. These also often prevent progress on specific issues where there may actually be a relatively high degree of potential agreement. Non-related international affairs can easily end up contaminating and unnecessarily politicizing the collective governance effort.

Collective governance tends to bypass the worst excesses of these ideological or parochial debates by focusing on a small set of undisputed technical requirements that rarely stir political machines. In this sense, the collective governance process can appear innately apolitical. Indeed, that it is the only basis under which companies would come to the table.

The EITI, for example, has never been formally negotiated by foreign ministry officials. Its international board does not have countries, companies and organizations represented with stated positions. It is individuals from countries, companies and organizations who serve on the board, albeit representing their respective countries and entities, and they therefore speak and act more freely. They feel mandated to speak to technical, rather than political, issues. In reality, as the process evolves, the lines blur.

It also helps that the EITI, like transparency efforts more generally, does not belong to any specific political philosophy—it is generally welcomed by promoters of social justice, economic growth, competition and equality.[2]

This also implies that, despite the demands of donors and others, time spent developing complex indicators of progress is also, at first, not time well spent. Any indicators should reflect the agreed products only. It is disappointing to would-be funders, but establishing expected outcomes and long-term goals risks opening up unhelpful debates and jeopardizing consensus.

The authors therefore conclude that there is limited benefit in a theory of change and little effort should be made to establish overarching goals. Instead, progress is reliant on establishing the clear reasons to come to the table, rather than grand common objectives. **Collective governance is more in the territory of mediation or conciliation than conventional development theory**. The main achievement at first is to get willingness to sit around the same table and agree on points of consensus and related actions. A relationship counsellor does not make assumptions about where the couple wishes the process to go, only that they talk about it. When negotiating peace between warring factions in Syria, the focus of the talks was not on any final solution or even ceasefire. Agreement on these aims was unthinkable. Instead, the talks focused on the more manageable, though considerably less ambitious, lower aim of agreeing humanitarian access to the besieged city of Homs.

The key is not necessarily to focus on the most important point of contention. Instead it is about how to find and move the point of consensus from

2   David Cameron's speech at the Open Government Partnership Summit in October 2013 is a particularly eloquent elaboration of the economic case for transparency (https://www.gov.uk/government/speeches/pm-speech-at-open-government-partnership-2013).

the narrow to the meaningful—iteration by iteration with no clear roadmap or time-frame. **How any collective governance effort evolves depends on what consensus can be gained at what time.**

The most obvious and common area for collective governance is the enforcement of codes and standards through quality reporting and a discussion of the implications of the reports. Alternatively, they might develop and analyse some sort of planning or forecasting model for the sector. These are shared immediate actions, but do not require a deeper agreement about the policy objectives or implications of the reporting or modelling.

From this narrow consensus on immediate actions, a wider set of collective actions can arise.

The EITI: progress despite lack of consensus on overarching objectives

The EITI has never had a full agreement about its most fundamental tenets. There are the EITI Principles (a statement of beliefs), but no agreed objectives. There was also, in due course, agreement on criteria that defined the scope of the EITI core activities, but this was an outcome, not an objective in its own right. To this day there remains a stark difference in basic EITI philosophy between those who see it as a centrally determined standard with well-defined requirements, and those that see it as a marketplace for country-led ideas with emphasis on in-country ownership. Trying to forge a single objective between these polar views was never considered necessary and the dichotomy has not been resolved. This has not prevented progress nor hampered the building of trust.

The EITI effectively broke these rules of development planning. Yet, to the credit of development partners, the process was accepted and is now hailed as a success by them. This story tells us a lot about the nature of collective governance as an adaptive, not planned, form of programming and has profound implications for development policy. Collective governance requires donors to be willing to take a walk in the dark.

5.1 Entry points: better to have the right people agreeing on a small thing than wrong people agreeing on the big thing

Finding the right starting point appears to have been the key design question for many collective governance efforts before and since EITI. It is the view of the authors that many make the fatal mistake of starting with the most important issue to achieve the result. It may seem counter-intuitive, but the governance entrepreneur must focus not on the important point, but on the point of most likely consensus: the point on which all parties can agree a way forward. That point of consensus might be seemingly irrelevant or mundane, but it is the entry point to get the key stakeholders, who might otherwise be throwing bricks at each other, around the table. More often than not, the most important point is too hot to touch. It is precisely the thing that the stakeholders are arguing about. Only once there is agreement on some relatively minor points rather than an important one, can progress be made. **As with any mediation, it is better to have the right people agreeing on a small thing than the wrong people agreeing on the big thing**.

In the case of EITI, the point of consensus was revenue transparency. For the Kimberly Process, it was traceability of diamonds. For the Voluntary Principles of Security in the Extractives Sector, it was protection of human rights. For the Ethical Trading Initiative, it was the promotion of workers' rights. They built vague principles around these points—often deliberately making them seem more important than they were in order to build a sense of purpose.

Processes around these principles built understanding and awareness which were then fed back to create specific rules for participation, compliance and impact. Then political and economic incentives were introduced (such as aid conditionality, correlation with foreign direct investment/sovereign debt rating/corruption perception/etc.) with a view to change behaviour.[3] Once momentum was established, consensus was consolidated, trust was built and then the conversation could broaden.

3   Carbonnier *et al.*, 2011.

5.2 The incremental pursuit of ambition

> Nobody made a greater mistake than he who did nothing because he could only do a little.
>
> Edmund Burke

The problem of governance is bigger than any one tool. It needs a toolbox. Don't oversell your collective governance process as the answer to all problems. It is only one tool in the toolbox. But just because it doesn't solve all problems doesn't mean that it solves none. The challenge is how to turn the tool from a screwdriver, which only deals with one issue, into a Swiss Army knife, that can adapt to many. The EITI was a tool for revenue transparency and has evolved into one that seeks to improve data about the management of the whole governance of the extractives chain from licence allocations to production, laws and fiscal arrangements, state ownership, social and sub-national payments, etc.

As with many international processes, the great temptation with the EITI was to create an all-encompassing solution to solve all the issues of oil, gas and mining governance. Not only would this have failed to attract the best people around the table, since there is no multi-stakeholder group that agrees on what to do on all these issues, but it would also have stepped on the toes of other complementary efforts. There were already initiatives for gas flaring, mineral tracking, public financial management, human rights, labour rights, environmental safeguarding, public–private partnership, contract negotiation, tax reform, indigenous peoples, and so on.

Instead of trying to find the supertool to fix all the problems, it is better to just try to find one small element that can be fixed collectively. Once progress has been made on this core product and trust built, you will need to find ways to incentivize wider innovation to make the process meaningful.

While the authors are establishing a framework for managing collective governance, they are not proponents of the concept of a "science of delivery" by which complex problems can be broken down into predictable, soluble parts, and spread through best practice. They believe that complex problems, such as the governance of natural resources, can only be solved by adaptation and iteration, and that collective governance is one forum available for such adaptation and iteration. That process begins by trying to agree an entry point.

Instead of a "science of delivery" the evolution of the EITI reflects the strength of **the incremental pursuit of ambition**.

The EITI: a platform for both focus and innovation

There is no magic bullet for most complex governance challenges, such as natural resource management. It would be naïve to think that any single intervention, including the EITI, could solve the violence, mismanagement and environmental disasters of the Niger Delta, for example. Yet the EITI's traditionally narrow focus on revenue transparency (up until the EITI Standard was launched in 2013) had been the chief criticism against it.

The common narrative was that the EITI was good at nailing implementing countries down to its minimum requirements, but had done less well in encouraging and incentivizing countries to go beyond. Consequently, progress was uneven.

As former EITI Board member and Global Witness campaigner, Diarmid O'Sullivan, put it: "A few strongly-performing [countries] followed by a long tail of countries where reports are late and of variable reliability. Some variation between countries is inevitable, maybe even necessary. But too small a head and too long a tail would make the EITI look like a peacock—a bird which is loud with its own importance, but more decorative than useful."[4]

The consensus approach of the EITI Standard had led to just one aspect—revenue transparency—being tackled first. Revenue transparency might not have been the most pressing or important issue in many resource-rich countries—it might have just been "cough medicine when I am dying of cancer" as one Tanzanian MP graphically once put it to Eddie—but it proved to be a sound entry point, indeed possibly the only entry point, for bringing all parties to the table.

In country after country, even basic revenue transparency became the starting point for other discussions. The EITI multi-stakeholder group created a powerful dynamic in which questions beyond "What is paid?" were increasingly asked, such as "What should be paid?", "Is it a good deal?", "How much is being made in volume and in profit?", "Where is it going?", "Who is benefiting?", etc.

The "global standard, local accountability" tag line and framework allowed progress and adaption. While the rules remained narrowly cast, the EITI framework was adapted by the countries to fit their local circumstances, such as Nigeria with production, Liberia in forestry and agriculture, Mongolia with licence transparency and environmental payments, Iraq with export sales and Kazakhstan with social payments. Contrary to popular belief, by 2013 there were very few EITI countries doing the bare minimum.

This increasing use of the EITI as a platform for further reform efforts was formalized in the EITI Standard in 2013, requiring more contextual information on size of sector; volume production; forecasts; payments and transfers to subnational levels of government; which government pot the money ends up in; licence transparency; the arrangements, sales and flows with

4   O'Sullivan, 2013.

state-owned companies; and transit payments. More detail was required on payments data, and contract transparency was encouraged. In short, countries were encouraged to shape their EITI process to the issues that they wanted discussed and addressed.

It would be naïve to think that the simple act of publishing data would always and sustainably improve management, just as it would be naïve to think that the multi-stakeholder platforms for dialogue would always sustainably improve management if not informed by good, timely data. There have been cases where the EITI's impact has been sadly limited by just focusing on one or the other. It was only when these elements of process (the collective governance) and product (the data) moved together that they have seen the EITI become a highly effective catalyst for wide-ranging reforms.

By focusing on the narrow consensus over the important, the process will likely come under serious criticism. The criticism should serve as a reminder of three things:

1. Keep the message about the role of the collective governance both clear and humble (see Chapter 6). The EITI Rules alone were necessary but not sufficient to tackle the challenge of resource management. To paraphrase Professor Paul Collier, "The EITI is the right place to start, but the wrong place to finish", though its collective governance platform was already a powerful tool for debate wider than revenue transparency.

2. Do not allow the process to become too siloed. It needs to link with other efforts. Its reports need to draw on other data and feed into the products of other processes.

3. Do not allow the process to stand still and expect it to survive. While focus is necessary to begin with, the process also needs to constantly push the boundaries. At the national level, it is easier to expand the process and move the consensus if the debate is already raging in the media and in the collective governance forum. At the international level, it is easier to get international movement if the process is country-owned and the countries are ahead of the international standard.

Getting things into perspective: what EITI compliance does and does not mean

The EITI shows that some kind of reward or feedback mechanism is necessary. For all the flaws of validation—the quality assurance mechanism to assess whether countries have achieved compliance—the EITI is unlikely to have been sustained without it. Compliance gives acknowledgement and encouragement (as does, to a lesser extent, candidature) and it is strongly fought for and prized. It keeps countries striving and engaged. The deadlines, as examined earlier, might be crude and occasionally counterproductive, but they have certainly made governments and their stakeholders deliver.

EITI compliance might be considered sacred but it should not be considered perfect. Compliance and other milestones give a win to the system that all stakeholders like. However, the country assessments on which they are based are technical assessments, not assessments of impact or political will. Nor does EITI compliance give the full picture on wider governance of the sector. It would clearly be a mistake anywhere to suggest that EITI compliance means "job done". Nigeria's compliance does not, of course, mean a clean bill of health for the oil and gas sector in Nigeria. It does not even mean that the governance of the sector in its totality is even improving. Some of the worst-offending countries no doubt implement the EITI successfully, while better governance performers remain stuck or outside the process.

A technical assessment is clearly not everything, but neither is it nothing. It is a starting point, not an end-point—a platform for change, rather than change in its own right. And that is significant and important.

Governments' capacities to assess and improve their revenue collection processes have increased. Nigeria's review of its first ten years of EITI reporting highlighted that at least US$9 billion more government revenue from the sector was received in 2008 than would have been the case before EITI reporting,[5] yet over the 2004–10 period absolute poverty rose by 6% to more than 60% of the population.

In the case of another controversial compliant country, Iraq, the governance implications of the EITI were, at the time of writing, limited, but the process was a good platform for public debate about important policy improvements. The fact that 34 buying companies based outside Iraq provided full disclosure for two years about how much they paid for oil, despite it not being in the contract nor being bound by Iraqi law, was remarkable, as well as paving a way for the debate about more potential for transparency in the trading of commodities. The report also helped identify many concerns. Chief among these was that the process showed that 690 million barrels were exported for US$52 billion in 2010, meaning that the opportunity cost for the 270 million barrels released for domestic use (predominantly electricity) was over

---

5   NEITI, 2012. The report is careful not to attribute direct causality between the implementation of the EITI and government revenue.

US$20 billion. That's the equivalent of around US$620 per Iraqi for electricity, compared with the US average consumption of US$729 per person. Yet in the United States, they consume almost seven times as much electricity per person than in Iraq.[6] This raised a big question not only on the use of the US$52 billion revenue, but also on the use of domestic oil.

The Democratic Republic of Congo (DRC) took a tortuous journey to compliance in 2014, but it was clear that the EITI has changed the culture in the sector from just extracting minerals to a greater focus on responsible management of the revenues for the benefit of the citizens of the DRC. This includes more focus on making sure that the government gets its due and that these revenues are adequately managed. Companies and government agencies have been systematically scrutinized, often for the first time. All contracts have been made available to the public and discussion on complex contractual agreements, including controversial barters, is taking place. It is no longer a taboo to discuss beneficial ownership, the flesh-and-bone owners of companies, and whether they are politically connected. Parliament and the media are discussing this issue openly. In a word, while significant challenges remain in the sector, trust has been built through the process.

The existence of EITI compliance inevitably leads to too much focus on "pass/fail" as the sole measure of the quality of EITI implementation. A binary standard was the right starting point and has generated a lot of good activity, but it is probably not the right end-point. There is the danger of "lock-in"— "compliance" is a difficult reward to shake off. There is a need to think more about nuanced assessments, recognition and incentivization as the EITI moves forward. Reward tools other than compliance will be needed that speak to the wider impact of the EITI.

The EITI is collectively thinking about how to provide incentives to countries that go beyond the standard. At present, the EITI chair provides awards to the best-performing countries. Some countries are highlighted more on the website, in progress reports and in good practice guidance. Some are asked to provide more training and peer guidance than others. However, all this stops short of a recognition within the EITI Standard itself.

The EITI: a note on timeliness of data

Through processes such as the Open Government Partnership, EITI, social media, WikiLeaks, etc., transparency and information disclosure is increasingly becoming the expectation, not the aspiration. However wide or small the focus of discussion, there is little point discussing out-of-date data. In 2009, Nigeria's most recent EITI data related to 2004. There was a wide scope of detailed

6   CIA, 2008.

and important information, but all the scandals and mismanagement that it revealed related to actors who had left the stage long ago.

The 2007 EITI report of the Democratic Republic of Congo was distributed in 2010. A Congolese woman told an EITI multi-stakeholder group member: "Your EITI is no use. It's like the doctor who arrives after the patient has died. What good is it?"

The EITI made a mistake in its first rules by not requiring timeliness of data. This was recognized and addressed. By the end of 2012, all EITI countries were required to have released data no more than two years old.

Several countries have begun to explore the potential for "real-time" EITI reporting. This is consistent with a broader shift towards EITI countries issuing several interlinked reports throughout the year, each focusing on different aspects of extractive industry management (such as licensing, contract transparency or beneficial ownership). This will lead to a more dynamic, relevant and interesting EITI process, and a steady stream of timely, comprehensive and reliable data, that can contribute to better-informed public debate.

5.3 Build in adaptability and learning

Even with the most brilliant governance entrepreneurial skills, it is bound to be difficult to gain backing, or even to see the potential for shifting the consensus. Collective governance needs to build in:

- Adaptability in the rules to avoid "lock-in" syndrome

- System refreshes through regular reviews and feedback mechanisms to push the boundaries and to refresh ideas

5.3.1 Avoiding lock-in syndrome

The complexity of the issues to be addressed by collective governance often makes rule-making challenging. Each case is unique and complex and cannot easily be captured in the rules. The 2013 standard acknowledged what was already the existing practice—that countries were allowed to present a case to the board to adapt the EITI Rules to their particular circumstance.[7]

---

7   Requirement 1.5 of the EITI Standard allows that if the country faces exceptional circumstances that necessitate deviation from the implementation requirements, it must seek prior EITI Board approval for adapted implementation. In

For example, with the authority of the board, Nigeria had undertaken ten years of revenue data on its massive and dominant oil and gas sector before it began coverage of the mineral sector. Similarly, with the board's authority, Iraq's reports covered export sales of oil and gas rather than tax, royalties and signature bonuses, because its oil and gas was entirely state owned and thus received its revenue only through sales. The United States has been allowed to proceed despite the fact that most oil, gas and mining activities occur on private land and therefore pay no rent or royalties to the federal government.

Finally, of course, the process has to flexible enough to adapt to changing circumstances. A country or organization that is compliant one moment might not be worthy of the title the next. In the EITI, for example, countries rich in oil, gas and minerals but weak in governance institutions tend to have a winner-takes-all political system and are especially vulnerable to coups. Since 2007, within EITI implementing countries, there have been coups or non-elected regime change in Burkina Faso, Central African Republic, Côte d'Ivoire, Guinea, Kyrgyzstan, Madagascar, Mali, Niger and Yemen. Many others have experienced instability that has temporarily prevented the functioning of collective government. As noted before, there have been incidents of imprisonment and harassment of civil society activists in many countries, and there has been an EITI Board debate over whether some of the requirements of the standard related to civil society have been breached in Azerbaijan. An international collective governance organization clearly needs to work out what to do in such circumstances, and the EITI has therefore established a Rapid Response Committee to make urgent recommendations to the board on a country's status.

There is an emerging debate on whether complex and expensive reconciliation processes, or even EITI reports, should always be required. The Norwegian EITI has produced six years of reports showing no discrepancy between company payments and government receipts. Does it continue to make sense to demand that they do this or could there be some sort of waiver or adaptation system?

Making the process adaptable is also a big element of making it more attractive to all potential users. That is critical for preventing unbalanced implementation which can be a threat for the entire enterprise. For example,

considering such requests, the EITI Board is required to place a priority on the need for comparable treatment between countries and ensure that the EITI Principles are upheld, including ensuring that the EITI process is sufficiently inclusive, and that the EITI report is comprehensive, reliable and will contribute to public debate (EITI, 2013a).

The EITI Rapid Response Committee

The EITI Rapid Response Committee addresses urgent issues related to in-country implementation on behalf of the EITI Board. It has dealt with coups, harassment of campaigners in the extractives sector, and draft NGO laws. The meetings have been irregular, short-noticed, unpredictable and often time-consuming.

The committee has occasionally made recommendations of suspension, but more often has proposed that the EITI Chair write to express the board's concerns and/or inviting the country to voluntarily suspend before the board imposes it on them.

governance problems are too often seen as belonging only to developing countries. Indeed most development interventions are exclusively designed for developed countries and donors to provide funding for interventions in developing countries. Yet the incentives for codes and standards for governance do not work like that. The problems they are trying to address are shared by rich and poor countries alike. In matters of corruption and human rights abuses, often the weak governance of a poor country has been exacerbated or exploited by the behaviour of rich countries, their companies and their consumers. Developing countries therefore understandably do not respond well to being preached at.

Furthermore, emerging economies, which often have the most pervasive forms of governance challenges, and also to some extent the less developed countries, do not aspire to join a club of what they might perceive as failed states. They want to join the OECD and the G20, and play by their rules and codes.

One can always employ the argument, "Don't look at other countries: if this is useful for you, do it." However, that argument only goes so far. Whether the complaint is real or an excuse for non-implementation, it needs to be addressed. The development movement is catching up with this. The financial collapse might have helped. Eddie went to a meeting of mining ministers in Addis Ababa in 2008 shortly after the financial collapse and heard one minister say to another: "So I see that they are nationalizing banks across America and Europe. That might be a good idea, but if we tried that they would never let us." Aggressive transfer pricing[8] has

---

8   In a 2007 survey assessing the economic practices of 476 multinational corporations, 80% acknowledge that transfer pricing remains central to their tax strategy (Tax Justice Network, 2009).

been taking place for years and significantly damaging developing country economies, yet only when it started to bite austerity-ridden Europe and the United States was concerted global action taken. So the sands have shifted. International efforts to tackle collective governance approaches must not only be implemented in developing countries.

Turning the EITI into a club for all those who can benefit from it

By the publication of this book in early 2015, the EITI still only had three OECD implementing countries: Norway, the UK and the United States. Unbalanced implementation risked confining EITI to what some potential implementers viewed as "a failed states club". There were many reasons for the unbalanced implementation. First, a missed opportunity by the UK government to set the example when establishing the EITI—a mistake not repeated by Norway when it became host to the International Secretariat nor, incidentally, by the UK when it later established the Construction Sector Transparency Initiative (CoST).

Second, the fact remained that the EITI was borne out of the "resource curse" issue—which was not recognized by many developed resource-rich countries, leaving them with the question: "What's the point of implementation for us?" Australia and the United States committed to pilot and implement the EITI respectively for domestic reasons in 2011 as a response to debates about "fair" tax returns. It therefore became clear that the EITI was no longer to be seen exclusively as a tool for addressing the resource curse. The commitment of the United States to implement was particularly ground-breaking. The UK, France, Germany and Italy began to ask whether they also might have domestic reasons for EITI implementation. They also committed to implement (or pilot in the case of Germany) around the G8 Summit in Lough Erne in June 2013. These efforts both to recognize their own country challenges, and to get their own "house in order", left no refuge for those who argued that they would not implement the EITI until it was a global standard.

Third, and perhaps most importantly, few OECD countries were actually resource rich. While the EITI seeks to be a global standard, there has never been a suggestion that all countries should implement. The EITI is only for countries with significant resource governance issues.

5.3.2 System refresh: using reviews and feedback mechanisms

The authors did not see the chance to develop the EITI as a revenue transparency mechanism. They held the criticisms, but were stuck in a mind-set that the narrow global focus allowed progress to happen in countries. What they did not see was that progress in countries allowed the global focus to become less narrow. It took the vision of Clare Short—backed up by an evaluation from Scanteam[9]—to see how the EITI minimum standard could be adapted beyond revenue transparency.

Furthermore, even if governance entrepreneurs can see the way forward, an independent evaluation or assessment that comes to the same conclusion can be a better channel than the entrepreneurs themselves to persuade stakeholders. If a governance entrepreneur cannot get buy-in and support for pushing the boundaries in a particular direction, commissioning an external evaluator can be more persuasive. Let others recommend the obvious way ahead if your advocacy might be mistaken for self-interest. It is hard for stakeholders to refute public and independent recommendations.

The authors have little new to offer on the literature on evaluations, peer review, research, technical assessments, and so on, but it is clear that such feedback loops need to be embedded in the design and development of a governance project such as the EITI. Even if they do not present new ideas, often external evaluations can be important tools for helping to move the consensus.

The flipside of that is, of course, that while regular evaluation and review is important, there is no substitute for real knowledge and understanding. Evaluations are an expensive business and a notoriously self-generating industry. Most of the best ideas should be able to come through the collective governance's own systems and people. Annual reviews of progress, peer reviews within the organization, management surveys, process audits, strategy retreats, and so on, are all lower cost ways of generating ideas from within the collective governance family. One of the strengths of the EITI is that the family is so broad that it contains skills on management, auditing, accounting, marketing, legal, technical skills related to the sector, etc. Individuals are asked to serve on the committees that best match their skills. On the other hand, few of them have much knowledge of EITI implementation. Whatever method is used and however it is embedded into the process, collective governance needs systematic refreshes to the system.

---

9   Scanteam, 2011.

The EITI and system refreshes

The EITI has been evaluated twice—in 2009 by Rainbow Consultants and in 2011 by Scanteam. The results of these, preliminary in the case of Scanteam, were presented to the conferences of those years. The Scanteam evaluation led to a two-year strategy review that culminated in the production of the 2013 EITI Standard. In 2014 Scanteam did an evaluation of the technical assistance provided to support the EITI process.

The conferences themselves create an opportunity for relaunching the EITI in different ways. For example, the 2011 theme was "Transparency Counts" and that of 2013 was "Beyond Transparency". All sorts of rebranding can take place at these major events.

The various committees of the board also undertake regular reviews—there have been at least two formal governance reviews, an expenditure review and a revenue review.

However, it is in the review of country work that the greatest potential exists for pushing the boundaries. The EITI country processes have an inbuilt planning, monitoring and evaluation system. Each country is required to have an EITI workplan updated annually explaining what it wishes to do with the process. It is then required to publish an annual report to monitor progress against its workplan. Then the three-yearly validation process does not just assess progress against the requirements of the Standard, but also an impact assessment of the process. Furthermore, each country EITI report contains recommendations for the process but these tend to focus only on the collection of data.

The authors believe that there is considerably more potential for country peer review or some sort of EITI expert panel to undertake a rolling review of EITI country processes and propose recommendations for how the processes can be made more relevant and meaningful in each country, and recommendations for the EITI global system as a whole. This is an issue that the EITI Board will come back to.

Lessons from chapter

- Focus on the possible, not the meaningful. The meaningful can come later.
- Hold your nerve against criticism of irrelevant focus, but make sure that the process is linked to other debates and processes.
- Keep challenging the stakeholders to move the consensus.
- Ensure that the rules are robust but do not lock in inflexibility.
- Build in systems to refresh the process, such as evaluation, strategy review, etc.

6
Getting the most from people

6.1 Win friends and influence people

> Paddington shouted "Help!", quietly at first so as not to disturb any-
> one.   Michael Bond, *A Bear Called Paddington*

Tailoring messages and using communication tools to build and sustain a
movement of support is one of the trickiest areas for a governance entrepreneur. As the quote above suggests, the trick is how to raise the alarm and get
action, without being accused of threatening any stakeholders' interests or
overselling the issue. The challenge exists at every stage:

- Persuading conflicting parties to the table while allowing sufficient
  space for conflict to rage to keep them there.

- Selling the process to garner support and momentum, while not
  overselling a narrow consensus.

- Even if the process is going smoothly, stakeholders might yet argue
  about the public policy implications of any product (such as a
  data report, a forecasting or planning report, etc.). The governance
  entrepreneur must promote the product while remaining neutral
  on its implications. This becomes more difficult the more meaningful the product becomes.

The communications challenge for collective governance is far greater
than the standard communications project which focuses on awareness
raising, campaigning, brand development, product promotion and dis-
semination. Instead the initial process is composed of three main elements:

1. Building support networks, getting endorsements and developing guidance—"co-ordinating energy".

2. Linking the process to wider debates, events and opportunities, and using the support network to take advantage—"strategic opportunism".

3. Then, once you have a product, such as data, make best use of it and get the support networks to analyse it and promote it to generate debate.

A governance entrepreneur should avoid the mistake of many development efforts, including to some extent the EITI, of overpromoting the process before having a product. In retrospect, it is surprising that the EITI was given so much leeway to promote itself as a transparency process before the end of 2008, when only about a dozen EITI reports had actually been published. Though perhaps the authors did not sufficiently acknowledge it at the time, it was a precarious moment for the EITI.

6.1.1 Co-ordinating energy: tailoring messages and building support networks

Real influence comes from an ability to build, use and expand networks. A governance entrepreneur will take a lot of time tailoring messages for different audiences to build a network. Collective governance is about a plurality of interests and views, often in sharp hostility towards one another. A simple adjustment of messages is not enough: a wholesale tailoring is required. The EITI quickly abandoned any idea of a standard presentation or briefing; what one audience wanted to hear about might bear no relationship to another. The language used by companies, for example, differs significantly from that of diplomats. The discussion in Chapter 5 is relevant here: different stakeholders are engaged for different reasons. There is therefore a need to appeal to those different reasons. To be effective here, a governance entrepreneur needs to get to know their stakeholders well, work out their motivation to come to the table and what they want from the process, tailor the message to them and then build the network. This is a process the authors call "co-ordinating energy".

Regular communications are also important. There are libraries full of advice to managers about stakeholder analysis and how to keep different types of stakeholder well informed. The authors do not claim to have much to add to that literature, except to emphasize the importance of keeping a two-way relationship going. While it is of course important to keep the stakeholders informed in a clever, tailored way, it is easy to overlook the importance to you of having these stakeholders feeling involved and engaged in the process. This should feel like the most exciting part of their job.

If possible, governance entrepreneurs will try to meet their stakeholders face to face. It will give the stakeholders confidence in the person and in the process, and allows for more open dialogue. You should consider inviting the stakeholders to your office so they get to meet the team and understand the process. Give them a sense of belonging and ownership of the process.

In addition, correspondence should not just be of a formal nature, such as newsletters, but also personal, such as passing on interesting and relevant news stories or asking the stakeholder's opinion about a specific topic. Pick up the phone. Get on Skype. Meet up for lunch. Almost every work dilemma that the governance entrepreneur will face should be generating many informal e-mails and calls to key stakeholders for opinion, advice and expertise. Where possible, credit them in speeches, in documentation, in conversation. Against our better instincts, we are all flattered to see our name in print or have a nod of appreciation our way. It may take time, but in essence a free workforce is being developed—many of whom are unencumbered by the neutrality constraints facing a governance entrepreneur.

You are no longer the lone wolf, but a spider in a web of others giving out the messages and creating sound and fury.

In the absence of any products at first, and later on perhaps in the absence of many meaningful products, all the governance entrepreneur will have at their disposal are endorsements and an explanation of the process. Endorsements and fact-sheets allow a governance entrepreneur to explain the process and to show that it has high-level support, while being humble and accurate about its demonstrated impact so far.

Getting good endorsements, however, is not straightforward. Again, having nurtured friends of the process, governance entrepreneurs need to draw on this support network to obtain big-name endorsements. Furthermore, the names have to be not only high profile, but also diverse enough to ensure a balance that pleases multiple audiences. The EITI spent years touting what was at first a somewhat feeble endorsement sheet to show that the process had traction, even if it did not yet have impact. With the enormous help of many friends of the process, who often went to extraordinary efforts, the endorsement sheet was increasingly more impressive. The EITI managed to get statements from heads of government, business leaders, prominent civil society activists, academics, journalists, etc. Even if statements from G8 and G20 summits were often somewhat hollow and even more difficult to get than any individual endorsement, they were significant of wider endorsement and they generated new interest from government officials, the media, businesses, campaigners and the public.

It is probably the case that collective governance efforts, such as the EITI, could make better use of social media to build co-ordinating energy. However, collective governance is not campaigning. The message cannot easily be captured in 140 characters. Social media does not allow much tailoring of messages. It is certainly useful for maintaining stakeholder interest and keeping a social network involved, but the authors are sceptical about its role as a network builder in the first place, except at the most superficial level.

The EITI abandoned the endorsement sheets in around 2010 in favour of stories of real impact. Perhaps even more than the endorsement sheets, these impact stories needed to be diverse and tailored to different audiences.

The EITI and UN General Assembly endorsement

In 2008, the authors spent what at the time seemed like a shocking proportion of time on a UN General Assembly Resolution noting support for the EITI. It was tabled by Azerbaijan and co-signed by 22 other countries, and eventually passed in September that year. Even though the statement was watered down through opposition by some prominent emerging economies, it set off a diplomatic process that raised the profile of the EITI in capitals all around the world, and left the EITI with an endorsement that could be used everywhere. Furthermore, it was one of the first steps in a long journey to demonstrate the EITI as a global process. Interestingly, while this international backing was obtained, it did not result in the EITI decision-making itself becoming more multilateral in its character.

EITI communication materials

Keeping the EITI family informed, engaged and enthused, across so many different interests, languages and methods, takes constant feeding of the monster. For the EITI, the production of communication tools and materials is a huge effort, involving the equivalent of at least four full-time staff, but it also pays enormous dividends. Of course, a newly formed or smaller collective governance process might not need nearly as much as the following list of regular EITI communications tools and materials, but this might be a useful guide to potential products.

- The EITI website: the first port of call for any casual interest, and also a library of information for highly engaged stakeholders.

- The EITI reports database which has access to all the EITI reports ever published, and a grand database that brings together a core of information for comparison and cross-referencing.

- Monthly newsletters (they used only to be every two months when they had less to say—an empty newsletter can be a turn-off).

- Confidential circulars to board members on average every two weeks, with a mix of issues for board decision, news and updates.

- Tailored circulars produced for different stakeholder groups, such as EITI national secretariats, companies, etc.

- The EITI annual report, which not only records the activities of the year, but also seeks to assess the impact and results of the process globally and across individual countries around the world.

- Finally, of course, the EITI has to make full use of social media. News items are posted on Facebook and Twitter, as other more informal communications—challenges, fun facts and questions.

On top of these regular communication tools, there are many other semi-static communication materials:

- EITI fact-sheets, and versions of these tailored for different audiences (e.g. Chinese, Middle East and North Africa, institutional investors, etc.), plus a checklist for countries planning to implement.

- The EITI Standard, including detailed guidance notes on every aspect of the standard.

- Guides for different audiences: business managers, parliamentarians, mining companies, etc.

- Leaflets on how to support the process—for countries, for companies, and for institutional investors.

- The EITI Articles of Association.

- Case studies.

Of course, almost all of these products need to be available in the EITI core languages of English, French and Russian, and many of them also in Spanish, Arabic and Portuguese. In addition, the International Secretariat develops many board papers and other correspondence with its key stakeholders.

Finally, the International Secretariat has promoted a number of competitions for EITI stakeholders in implementing countries to generate both interest and use of the EITI process and communication tools for ourselves. They have held a **video competition** to tell the EITI story from the perspective of the people on the ground, which was broadcast at the Sydney conference. They held an **infographic competition** to promote the use of EITI data, which was used to explain the results of the EITI process elsewhere. Finally, they held two **photo competitions** which again promoted visual storytelling, as well as providing lots of material for the EITI annual report and presentations.

There are only two board members remaining from the original group in 2007. Most board seats and representatives from governments, companies or civil society have changed hands at least four times in that time. It is remarkable how much support networks rely on individuals rather than institutions. Turnover of stakeholders means starting a relationship from scratch. It is a constant challenge to keep the EITI relevant and exciting to each new generation of stakeholders. The flipside is that it is enormously rewarding to "recruit" a new friend of the EITI who goes on to transform their organization's interest, engagement and support of the process. This book acknowledges their contributions in the leadership section. Furthermore, the friends that leave the process are rarely totally lost to the EITI—they tend to go on to promote the process wherever they are, often appearing in unlikely places as rogue EITI ambassadors in helpful forums. Slowly, the EITI process has become institutionalized into these organizations. Rightly or wrongly, for example, countries and development partners on occasion agree EITI implementation as an aid conditionality benchmark.[1]

6.1.2 Strategic opportunism

> There is a tide in the affairs of men.
> Which, taken at the flood, leads on to fortune;
> Omitted, all the voyage of their life
> Is bound in shallows and in miseries.
> On such a full sea are we now afloat,
> And we must take the current when it serves,
> Or lose our ventures.
> William Shakespeare, *Julius Caesar*, Act 4, Scene 3, 218–24

Governance entrepreneurs do not just need to be able to prepare for the weather, they need to make the weather. Opportunities come to those who invite them, not to those who wait for them—a process the authors call

---

[1] David-Barrett and Okamura, 2012.

"strategic opportunism". This relies heavily on having built your network of support—your spider's web. A governance entrepreneur looks ahead, sees the opportunities, and draws on that network of support. Having built their networks, governance entrepreneurs seize opportunities by:

- Increasingly turning from joining other people's debates, forums and conferences, to facilitating their own. Part of building a support network is agreeing to participate in a lot of events. Some of these will be useful in their own right, more of them will be useful just for building support from key stakeholders, i.e. doing them a favour, getting more opportunities to build face-to-face contact with them and their networks, and being able to praise them in front of their bosses. These are important credits in the account. However, roving around the conferences and events of others does not set the agenda. At best, you help manage the agendas of others by influencing the content and messages of their events. There will come a time to cash in the credits and draw the big names to your event, with your agenda and with your messages. The tendency in international development and politics to respond positively to invitations to address conferences must be balanced with a more active mind-set of planning yourself where you need to be. This can be challenging, as it may often result in travelling far to an unpromising outreach meeting with uncertain outcomes, something that can feel less secure than going to a long-planned big-set conference.

- Keep abreast of emerging scenario analyses and then try to build a resilient strategy for staying relevant in all scenarios. Scenario analysis has become a well-used tool for strategic management. It is particularly relevant in global organizations. The strength of such modelling is not so much to forecast the future, but to outline different scenarios of the future and to adapt the collective governance process to remain relevant in all geopolitical situations. The authors note that the language of "resilience" is taking over from "sustainability" in much development practice. This fits well with the increasing turbulence of global politics.

- Spend a lot of time looking for opportunities, such as G20 summits, high-profile visits to implementing and potentially implementing countries, upcoming international or global reports, etc. To paraphrase Oscar Wilde, there is one thing worse than being spoken ill of, and that is not being spoken about at all. When these

The EITI global conferences

The EITI has held six global conferences—each one bigger than the previous in participants, profile, and ambition. The 2013 conference in Sydney had over a thousand participants from more than 80 countries. In a world of too many huge, expensive, self-promoting and time-consuming development conferences, some stern questions have to be asked about their value. There are three key reasons for continuing to hold them:

1. To link the EITI with wider geopolitics: recent conferences have been held amid global events such as WikiLeaks, uprisings in North Africa, and fluctuating commodity prices. The participation of so many prominent and interesting figures helped draw out the linkages.

2. To provide a forum for all key stakeholders to debate contentious issues in the sector: recent conferences have provided a forum for debate about mandatory disclosure laws in the oil, gas and mining sectors.

3. To ground the global EITI in national processes: through the EITI national exhibitions, implementing countries effectively compete with, and inspire, one another to demonstrate products and progress.

However, the main benefit of the conferences has been to act as a catalyst for action and a forum for building support. The conferences provide a milestone by which to complete reports and publications. They lead to a flurry of briefings for inclusion in presidents' or ministers' speaking points. They enable the convening of meetings with key leaders, activists and business people, and become a reminder of how much EITI matters to their agendas. EITI national bodies get a higher profile in their country and possibly more resources and political support. The EITI globally gets a window of opportunity to engage with new countries, companies, media and key individual champions.

high-profile opportunities come, call up your networks to get on to the agenda, contribute tailored briefings, write a box in a report or a blog in the media. Not all of these efforts will be fruitful, but nine fruitless efforts are worth one mention in a G20 declaration or an endorsement from a prime minister. This is not the same as jumping at every opportunity. Looking for strategic opportunities is a matter of assessing opportunity cost, and thus is as much about turning down invitations as it is about accepting.

The EITI and strategic opportunism

> Luck is when preparation meets opportunity.
> Lucius Annaeus Seneca, Roman philosopher

The EITI's rise did not happen in isolation. It rode on the crest of some important geopolitical waves which the EITI's stakeholders did their best to catch and to sustain. Below are the key waves with a short explanation about how the EITI positioned itself to take advantage. The last one is about collective governance. A book could probably be written about the EITI and the irresistible rise of any of these trends.

- "Transparency" and "accountability" have become central tenets of all governance discussions. Back in 2007, it was spoken about mainly as a flipside to corruption, but it has grown across all governance issues. Barely an international meeting takes place on any topic without a session and a declaration on transparency. The right to information is increasingly considered a universal human right. This has happened on the back of a financial collapse that many believed was stoked by secrecy and opacity. The Arab Spring led to increased demands for accountability of government. The WikiLeaks debate has changed the nature of government relations with each other and the public. Much governance debate has shifted from "need to know" to an assumption of openness unless there is a strong case for confidentiality. Transparency has become the expectation, not the aspiration. The EITI has promoted itself as a tangible process that not only ensures transparency, but also links the release of information to an accountability forum. The global governance processes have been only too willing to grab a tangible process to put action alongside their words.

- Governance of extractive industries: in 2011, the musician and activist, Bono, echoed Mo Ibrahim's comment that better governance in the sector "is bigger than debt cancellation for Africa".[2] Since then, commodities prices have fallen but the governance of the extractives remains high on the priority list. In many countries, such as Uganda, Tanzania and Mozambique, there is probably only one more electoral cycle in which aid will be the biggest source of government revenue. Revenue from oil, gas and mining will take over. It has already done so in Ghana. Development partners are looking for tools other than aid to ensure good governance. The EITI gives them a tool and has been seized on by many in the development community for its potential. More cynically some argue that Western governments, concerned that their extractive businesses in developing countries are losing out to state-owned companies and companies from the East, especially China, are finally becoming interested in improving the rules of the game to help their companies keep their competitive advantage.

- Tax avoidance issues: Professor Paul Collier regularly gives an ironic note of gratitude to Starbucks, Amazon and Google for putting tax issues so high up the development agenda, leading to it (alongside transparency and trade) being the focus of 2013 G8 Summit. Those companies were in the centre of a media storm for paying so little tax in the UK, claiming that they were making operating losses there despite massive global profits and expanding UK operations. This alleged exporting of profits from tax-high to tax-low environments, or transfer mispricing as it is commonly known, has been going on for years and is, on the whole, legal. Governments in both developed and developing countries have been victims but, of course, it has disproportionately affected poorer countries where such taxes make up a greater proportion of potential government revenue. Between 2008 and 2010, transfer mispricing cost Africa an average US$38.4 billion every year, more than its inflows from either international aid or foreign direct investment.[3] The Netherlands has one tax official for every 333 people compared with Nigeria where there is one for every 28,000.[4] It is surely no coincidence that it was during a period of austerity that developed countries decided to crack down too. However, there are a few information tools to detect how much should be paid in each country. While the EITI does not provide a method for determining profit in a jurisdiction, it can provide information on production, on exports, on sales prices, on costs of production, on contracts and tax regime, and on how much was actually paid. A quick analysis of these figures, as has been done in Zambia and Tanzania, soon shows if there is anything clearly out of kilter and can require the company to justify why that should be.

• The era of Big Data and the rise of indices, and their ability to use and confuse: there has been a recent explosion of data availability, which is especially true of the oil sector. Geological data that had been gathered at a cost of US$15 million was posted on the Internet by Nova Scotia in order to encourage investors, who eventually committed US$900 million.[5] Project-by-project data, alongside costs of production, will bring us closer than ever before to where oil companies make profits. The Natural Resource Governance Institute is using a complex index to rank countries on their governance of the extractives sector.[6] Whether you belong to the view that this data will do more to confuse or to inform, it is clear that it is being increasingly demanded, increasingly used and increasingly linked up than ever before. Furthermore, there is an increasing demand not just for the data, but for the "meta-data", i.e. the ability to mine it for meaning and context. The EITI has positioned itself well to require reliable data agreed by a wide range of stakeholders, underpinned by context and commentary, comparable across years and across countries, and coded to be interoperable with other data sets.

3   Africa Progress Panel, 2013: 66.
4   Forbes, 2013.
5   OpenOil, 2012: 3.
6   Natural Resource Governance Institute, 2014.

- Codes and standards, rather than global laws, to encourage good governance: Professor Paul Collier[7] and others have argued that, in the absence of effective global governance, nationally applied codes and standards have established good practice and prevented many of the worst aspects of the race to the bottom. Processes such as the Open Government Partnership have incentivized countries to bring these efforts to the international table, and charters such as Collier's own Natural Resource Charter[8] have established policy blueprints. The EITI is, of course, a global standard applied nationally.

- And, of course, collective action solutions to intractable governance problems: see Chapter 1 on the irresistible rise of collective governance.

In summary, the EITI arose and was sustained in a perfect storm to which, the authors believe, the EITI has modestly contributed. On the whole, this was achieved by subtle management rather than major grandstanding. If there was a big conference on energy security, the EITI International Secretariat would call up the speakers linking the EITI to energy security. If a major leader was visiting an important country, they would contact their office and provide a tailored brief on the EITI for that country. If a country was presiding over the G20, they would make contact with their liaison person in that government a year in advance to ensure that the EITI was on the agenda. If there was a BP oil spillage in the Gulf of Mexico, they would write an op-ed relating the EITI to managing risk and publish it in the relevant sympathetic media. All this tailoring took a long time and bore mixed fruits, but without having previously nurtured a supportive network in each lead organization it would have come to little.

6.1.3 Focus on the product not the process: use the data to stimulate public debate

While endorsements, explanations, building support networks and providing co-ordinating energy are essential and time-consuming components of a collective governance process, the authors would advise a measured external communications effort until there is something substantive to say. The collective governance process is only a means to an end. Its product— be it data, narrative reports, or other—must be demanded, disseminated, understood, used and debated. This is when governance entrepreneurs must again draw on their support networks. They should get the appropriate stories to the appropriate stakeholder group, and then stimulate, nudge and facilitate them to get a balanced and informed public debate going. It

7   Collier, 2007.
8   See www.naturalresourcecharter.org.

is the public debate, not the process, that should be the focus of external communication efforts.

The EITI and parliamentarians, the media and academics

The EITI engaged parliamentarians, the media and academics in its international and country processes. While they were happy to give endorsements, they were reluctant to join governance processes such as the national EITI commissions. In retrospect, it is clear that these stakeholders, unlike others, were only likely to be interested in the product (i.e. the data) rather than the process (i.e. the national commission). The EITI should have realized that they would only get involved when they saw what the product offered them: information about how their government's share of revenue from its commodities compared to that of other countries, how licences were allocated in their country, whether the contracts were transparent and followed, how the state-owned company related to the ministry, unpaid revenue, and so on. Since there has been a critical mass of EITI reports of good quality across a number of countries with more contextual information, these groups have become more engaged. With the EITI increasingly categorized with IMF coding, more interoperability is possible with other data—budgetary, population and services—and more analysis across times, countries and regions. Crucially also for parliamentarians and for the media, timeliness of data has been critical.

Once there is something to say, a governance entrepreneur can start spending less time on gaining endorsements and more time on stimulating demand and interest in the product. At the country level, this needs significant effort to make the message available nationally. There is a lot of literature on development communications to which the authors have little to add, but the following boxes outline their experiences.

The challenge for the governance entrepreneur becomes how to tell an interesting and important story while remaining neutral to political interests. The way of telling the story is, in itself, a tool for moving the consensus. A skilful governance entrepreneur will bring a story of shared data to life by suggesting areas where the data leads to an agreed analysis, and agreed areas requiring more information. It is not for the collective governance group to agree solutions.

The EITI and telling the story

Despite strong anecdotal evidence of its success, the EITI suffers from the same challenge as most governance measures—how to establish whether it really does lead to better natural resource governance, less corruption, more accountability and, ultimately, to more citizens in more resource-rich countries reaping more benefits from their wealth. That will only happen when data is understood, used and communicated.

When an EITI report is published, the International Secretariat analyses the report, tries to pull out the interesting facts and story, and then publishes a news item for the EITI website which seeks to present the interesting facts in a balanced way.

This type of use of data is a key challenge. Successes with efforts such as the EITI will be false victories unless civil society actors—the media, the communities, and so on—know how to use the information that is becoming available. The products of transparency must be used, analysed and turned into public debate.

Even the narrow EITI processes can reveal shocking revelations. The 2008 and 2009 EITI reports from the Democratic Republic of Congo showed that the government received less than US$200 million for its mining resources. That's less than US$1 per person per year for resources that are linked to the deaths of over five million people and have thrown the economy back many generations. That involved no disaggregation, no contract transparency, no discussion of subnational transfers, nor the role of the state-owned companies—it was a stark aggregated figure that, through good telling of the story, rightly raised the debate.

In 2010, Mongolia produced a thousand-page EITI report. While the report contained important and startling information about the sector, it was impenetrable and uninviting, and barely contributed to public debate. Most reports are becoming more readable but, given the technicalities of the subject, they often need to be accompanied by shorter "popular" versions with the key facts and figures. Often more useful are infographics: these are graphic representations of information, data or knowledge, which are intended to quickly and clearly present complex information, patterns and trends (see overleaf an example from the Oil Revenue Tracking Initiative in Nigeria, courtesy of the Yar'Adua Foundation). Infographics or other important messages can also be provided through cartoons or animation. But all of these are static communication tools. Affected communities often have high levels of illiteracy and the EITI has benefited from "info-mediaries" or "data tellers". These are members of the community who explain the data in simple, stark language and then stimulate and facilitate debate. They do this directly or through radio, theatre, etc.

Building on this, reports could increasingly be live online documents perhaps in the form of a portal to other information (i.e. an EITI website that simply links the reader to other databases, perhaps with a small narrative or

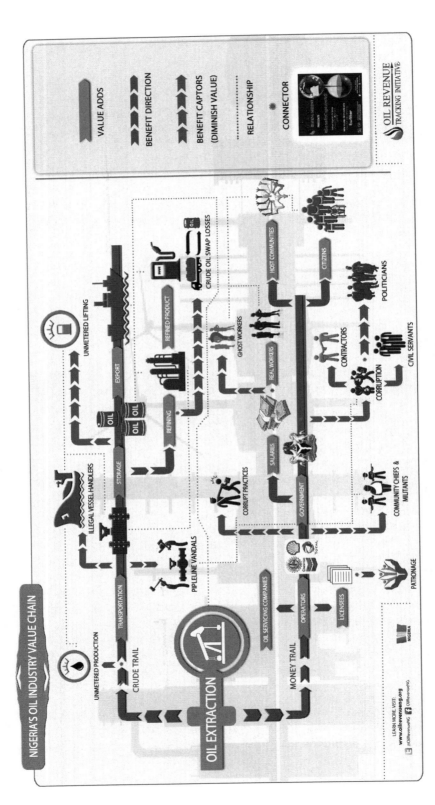

NIGERIA'S OIL INDUSTRY VALUE CHAIN

VALUE ADDS

BENEFIT DIRECTION

BENEFIT CAPTORS
(DIMINISH VALUE)

RELATIONSHIP

CONNECTOR

OIL REVENUE
TRACKING INITIATIVE

OIL EXTRACTION

UNMETERED PRODUCTION

CRUDE TRAIL

TRANSPORTATION

STORAGE

EXPORT

UNMETERED LIFTING

ILLEGAL VESSEL HANDLERS

PIPELINE VANDALS

REFINING

REFINED PRODUCT

CRUDE OIL SWAP LOSSES

MONEY TRAIL

OIL SERVICING COMPANIES

OPERATORS

LICENSEES

PATRONAGE

CORRUPT PRACTICES

SALARIES

GOVERNMENT

GHOST WORKERS

REAL WORKERS

CORRUPTION

HOST COMMUNITIES

CITIZENS

CONTRACTORS

POLITICIANS

CIVIL SERVANTS

COMMUNITY CHIEFS &
MILITANTS

LEARN MORE. visit:
www.oilrevenueng.org
@OilRevenueNG   OilRevenueNG

NIGERIA

visual commentary on how the sector is managed), or a map showing where the resources have been explored and/or extracted, with clickable functions to provide more detailed information on licences, contracts, ownership, production, forecasts, taxes/royalties/fees and social payments.

It is shocking how little reliable information there is at the global level about the oil, gas and mining sector given its governance implications. The secretariat has also developed an easy-to-use database to compare EITI data across time and across countries. But there is much more data out there which will grow rapidly with the introduction of mandatory disclosure laws in the United States and European Union, and with Open Government Partnership initiatives. The secretariat works with other institutions, such as the IMF, to codify its data and to develop a global database for the sector that links the EITI information to other relevant data sets, so that national and international policy-making might be better informed.

Finally, the secretariat is working with researchers to explore the wider impacts of EITI implementation, such as whether it reduces perceptions of corruption[9] and increases foreign direct investment.[10]

The Nigerian EITI and FOSTER: what's the point of transparency?

Accountability needs both public debate and a mechanism for holding the decision-makers to account. A well-designed collective governance programme works on both of these points. In Nigeria, the Facility for Oil Sector Transparency and Reform in Nigeria (FOSTER) programme strengthened bodies to collect the data (the Nigerian EITI), to use and debate it (the National Assembly, the Nigerian EITI and a wide range of civil society and accountability actors including petroleum unions, business associations, media institutions, research think-tanks and advocacy groups), and to act on it (the Ministry of Finance, the upstream and downstream industry regulators, and the oil spill response agency).

Interventions focused on:

- Increasing government revenue both as a proportion of output and by encouraging greater investment and production in the industry overall.

- Reducing the amount of revenue lost through rent-seeking, leakages and theft.

9   David-Barrett and Okamura, 2012.
10   Schmaljohann, 2013. By using a sample of 81 countries, this research found that countries joining the EITI have increased inflows of foreign direct investment. It concludes that investors see the EITI commitment as a signal of trustworthy willingness to reform, and to increase transparency and accountability in the extractive sector.

- Increasing fiscal discipline, performance and accountability through better public financial management and saving of oil revenue.

Like EITI, FOSTER worked iteratively with the key political forces by building relationships to support change, and worked to shift incentives and dynamics for successful reform. It took a venture capital approach to identify and create opportunities for supporting change, to take risks and to avoid early setting of objectives and long-term commitment of resources. A wide range of short-term projects allowed the programme to be flexible and responsive to early successes with further investment, building credibility and maintaining pressure on actors to deliver tangible results.

The same support is needed internationally to support the data-collection body (the EITI International Secretariat), the bodies that use the data (especially the Publish What You Pay coalition), and the bodies that can act on it (the implementing countries).

6.2 Leadership and the heroism of quiet compromise

There is not much new to be said about leadership. Like pornography, it is difficult to define, but you know when you have seen it. It comes in many forms beyond the classic spearheading an organization and commanding a high profile. Here we consider the leadership skills particularly relevant in collective governance. It is those more mundane, quieter (but every bit as heroic) skills that are required for effective collective governance.

In the film *Lincoln* (2012), Tommy Lee Jones plays a radical Republican congressman, Thaddeus Stevens of Pennsylvania. In the film, Stevens is a fervent anti-discrimination radical who does not trust Lincoln's intentions or methods to pass what Stevens considers the half measure of an amendment to abolish slavery. Even though he remains unsure of the outcome, disgusted by the method of "persuasion", irritated by the pace of progress, embittered by personal rivalry, and unsatisfied with the proposal, Stevens moderates his position on racial equity in the debate in the House of Representatives. This ensures that the amendment passes.

Whether or not the story is exaggerated and dramatized for the film, what is depicted is rare in Hollywood, where leadership is usually portrayed by an uncompromising morally driven visionary who succeeds against all odds. Instead, Stevens is portrayed as a man of principle who heroically agrees to compromise and reduce his ambitions in order to take a step in the right direction. Lincoln, on the other hand, is portrayed as a canny pragmatist who seeks an incremental route to a visionary outcome. They come to a qualified respect of one another. In reality, Lincoln described Stevens and his fellow radicals as "the unhandiest devils in the world to deal with [but] with their faces ... set Zionwards".[11] For his part, Stevens stated that "the greatest measure of the nineteenth century was passed by corruption, aided and abetted by the purest man in America".[12]

Though perhaps of less historical import or dramatic tension, the authors recognize these moral and practical struggles as the everyday reality of the EITI. This heroism of quiet compromise is the most overlooked of all the leadership skills, yet the most in demand for success in collective governance. The history of the EITI would have been short and brutal without an extraordinary number of these heroes.

The leadership skills required of stakeholders are varied, and even if all stakeholders do not necessarily possess all of them, they must exist among the collective governance group somewhere. What follows is an indicative, rather than exhaustive, list of the skills, institutions and individuals that were needed to get the EITI process functioning at the international level:[13]

- **Innovator**, to see the issue and outline the opportunity: as previously noted, much credit here to Global Witness and Publish What You Pay.

- **Initiator**, to get other stakeholders to see the opportunity for collective governance: PWYP and the then Revenue Watch Institute for civil society, BP and Shell for oil and gas companies, the International Council for Metals and Minerals for mining companies, F&C Asset Management for institutional investors, and the UK government for governments.

- **Convener**, to bring the disparate stakeholders to the table: the UK government, though each constituency also had their conveners as above.

- **Talent-spotter**, to assess who are going to be the key stakeholders for each particular issue, i.e. the "horses for courses": the

11   Herbert, 2004.
12   Scovel, 1898.
13   The authors are grateful to Chris Ansell and Alison Gash, whose "Collaborative governance in theory and practice" (Ansell and Gash, 2008) prompted some of the suggestions on this list.

EITI International Advisory Group (forerunner to the EITI Board) was somewhat self-selecting, but the UK government did well to approach Peter Eigen to chair, and the board did well to approach Clare Short.

- **Facilitator and mediator**, to steer the stakeholders to a position of consensus: Peter Eigen and then Clare Short.

- **Risk-taker**, to be willing to operate differently with conflicting stakeholders, to act outside personal and institutional comfort zones and conventional job descriptions, to put controversial proposals on the table, and to push boundaries: extraordinary examples from all sides.

- **Negotiator**, to argue positions, but be willing to compromise, and to take along followers, but continue to work with opponents: ditto.

- **Arbiter**, to judge between discussions and set out a way forward: Peter Eigen and Clare Short.

- **Adviser/counsellor**, to listen to concerns and suggest ways forward: the EITI benefited from numerous wise old heads inside and outside the process.

- **Catalyst**, to keep the energy going and the consensus moving: foremost credit to tireless campaigning by NGOs and journalists on natural resource governance, even during dark days when the issue appeared to be either too complex or too low profile for traction.

Of course, the list could go on. Perhaps, however, like Thaddeus Stevens, the most underrated skill was a willingness to make big decisions with a low or diminished profile. Collective governance, like peacemaking, is not about grandstanding or claiming individual wins. Leadership is about knowing when to give your opponents little gains. This is not just something that the head of an organization might do, but the ordinary middle people who are charged with making the progress. Powerful people accept the need to give over power to those who oppose them in order to address previous power imbalances.

Such individuals are likely to come to the table with more likelihood of being berated for conceding on key issues than being congratulated for averting further conflict and costs. They are usually taking big personal and career risks and must be determined to put long-term outcomes ahead of short-term gains. In many ways, engagement with collective governance is, for the stakeholders, something of a thankless task. Because they are unlikely to receive much recognition for their work within their institutions, it is even more important for the governance entrepreneur to find ways to give recognition for efforts, to build personal relationships and to excite individual action.

Furthermore, there have been many cases in resource-rich countries where individuals have put themselves or others in direct personal danger for the EITI cause. Some stakeholders have been imprisoned for their activism. Some have received death threats and unfounded career-threatening accusations. Some have been bullied, shunned or overlooked. Where undeserved vested interests are threatened, violence is often close behind. While these cases are too sensitive to outline here, the authors hope that this chapter can be a public tribute to those who have stood steadfast against these actions, threats and accusations. These are the brave unnamed soldiers on whose shoulders the EITI has made progress.

Governance entrepreneurs themselves will have to demonstrate leadership skills, yet should not seek nor expect credit in the short term as success of collective governance relies on the credit largely being taken by stakeholders, not the facilitator. Many of the skills of a governance entrepreneur have been highlighted elsewhere in this book (innovative diplomacy, co-ordinating energy, entrepreneurial risk-taking, communicator, etc.). The most important leadership skill will be to bring out the leadership skills in others. This involves understanding the motivations of each stakeholder and to create the best environment to get the most from each one. It is, of course, obvious but must not be overlooked—the governance entrepreneur needs to be a strong manager in order to oversee the mechanics of the process.

The EITI and the story of exceptional leadership

For the EITI to succeed, exceptional leadership has been required at all levels. At the highest level and in the beginning, individual leaders from civil society, companies and governments came to the conclusion that they would have to work differently to resolve the challenge of governing natural resources. The financier, George Soros, saw the need to focus and fund campaigning on the unfashionable subject of natural resource transparency. John Browne, the CEO of BP, did not take simple refuge in business inertia and agreed that the demands of the campaigners should be heeded. Olusegun Obasanjo, the President of Nigeria and his team of Ngozi Okonjo-Iweala and Obiageli Ezekwesili, committed Nigeria to pilot the process in a brave bid to reform the sector.

At the level of the International Advisory Group, individuals such as Gavin Hayman of Global Witness recognized the need to combine campaigning with dialogue. Karina Litvack of F&C Asset Management saw the role of institutional investors to lift the horizons of the big extractive companies and urge more collaboration around new business models. On the oil side, John Kelly (ExxonMobil), Alan Detheridge (Shell) and Stuart Brooks (ChevronTexaco) engaged immediately on the idea of a new model for transparency and partnership. On the mining side, Edward Bickham (AngloAmerican), Paul Mitchell (ICMM) and Shaun Stewart (Rio Tinto) were among those who recognized the opportunity for explaining more about the contribution of mining to the economy and society as a whole. Shahmar Movsumov, head of the Azerbaijan State Oil Fund, sought strong political backing, and defined and drove the Azeri process that shaped so much of the global process. Professor Humphrey Asobie worked with Obiageli Ezekwesili to develop a different type of model for Nigeria which showed the initiative's potential. DFID's convening power, originally under Clare Short, and with the management of Ben Mellor, gave the process political life at the international level. There were, of course, many more who contributed in extraordinary ways and are too numerous to mention here. What these and others did was to show enough creativity and vision to be willing to break out of their narrow interests and comfort zones.

The EITI has also been blessed with two exceptional chairs who demonstrated the right skills for the right moment in the development of the process. Peter Eigen came to the process in 2005 and steered it with consummate consensus-building skill, unique determination and boundless energy. In 2011, he handed a very healthy global organization over to Clare Short, who brought her extraordinary vision and political skills to drive through a complex and ambitious strategy review in a way that few others could have done, let alone even envisaged.

Lessons from chapter

- Recognize your role as a spider in a web. Build a support network by meeting face to face with potential friends of the process—find out their motivations and tailor your messages.

- In the absence of results, get endorsements and explain the process.

- Look for opportunities and make sure that you have enough friends in the right places to take advantage of them.

- Use support networks to make best use of the product to generate debate.

- Don't oversell the process. Sell the product.

- Help support networks to use the data, turn it into stories and stimulate analysis and debate.

- Seek out leadership skills at all levels.

- Embrace and acknowledge the heroism of compromise through documentation, speeches and discussion.

7
Governing the governance

This chapter is about the design of the institutions and rules of collective governance. It starts by trying to unpick the tension between the fact that there are too many institutions in development and in governance, and yet collective governance needs institutions in order to survive and develop. Collective solutions to governance challenges need to be carefully designed and calibrated to avoid ineffective bureaucracies failing to exceed a lowest common denominator of limited value. But, together with other governance efforts, voluntary as well as mandatory, these institutions can form part of the battery of efforts required to ensure good governance and to take policy where it has not been before.

The authors would advocate for a MAD approach to this "governing the governance"—mandate, accountability and demand:

- Mandate: build fit for purpose, not fit for themselves, institutions

- Accountability: build accountability *alongside* capabilities

- Demand: ensure that the institutions are and continue to be demand-driven

Following this structure, this chapter will explore the role of these institutions, their systems of accountability and some of the day-to-day challenges faced in managing them, including establishing funding arrangements, codes of conduct, reward mechanisms, adaptability of rules, balancing being a global standard and a national process, and avoiding having the standard used for whitewash.

As in previous chapters, the key message is that the governance arrangements for collective governance processes should not be established rigidly from the beginning, but need to be able to evolve as the organization grows and adapts. Then, in Chapter 8, the authors conclude with some guidance on how to assess when the institutions have achieved their purpose and should close.

7.1 Mandate: build institutions that are fit for purpose, not fit for themselves

Given that the authors define collective governance by having to have formal institutions, collective governance, by definition, needs to work out its institution building. A central multi-stakeholder body has to be established at some point. Furthermore, such a body may need some executive arm or secretariat to arrange its work. The EITI needed at least a secretariat to arrange meetings and papers. It seems likely that the lack of a secretariat and weak structures have contributed to some of the challenges experienced by the Kimberly Process Certification Scheme. However, the authors would caution against the "big bang" approach of setting up such bodies. They should only be established once the demand is clear.

Just because collective governance needs its own dedicated organization, it does not mean that the organization has to be the starting point—the institutions should only be formalized once the process has momentum. As with the arrangements for objectives, representation, legislative underpinning, and so on, the authors would advise that as little time as possible is spent early on defining and formalizing the institutions of collective governance. Until trust has been established, too much definition and too much formality will simply be another point of contention.

In building institutions involving different stakeholders, it must be taken into account that different stakeholders work at different speeds and with different priorities. In general, government representatives appear to have greater stamina for drawn-out conversations about process, protocol and procedures. Company representatives, on the other hand, appear eager to avoid creating organizations that may engage in empire building and long-term funding commitments, and be more interested in an early focus on concrete progress. Lastly, civil society representatives, fearful that companies and governments will water down progress and commitments, seem generally interested in ensuring safeguards of their influence.

Usually, institutions can start off on an informal basis. The EITI had an International Advisory Group rather than a formal board until 2007, and most of the EITI national commissions have limited legal or formal status: they have just been set up on the instruction of the government. Secretariats can incubate inside other institutions early on, so long as dedicated resources and incentives are provided. The EITI International Secretariat was hosted by the UK government's Department for International Development until 2007, and most of the national secretariats reside within government ministries. This also postpones big areas of contention, such as funding the institutions and legal entity, until parties are agreed that such bodies need to be formalized.

It is not easy to manage a shoestring operation run out of a cupboard in a wider organization—resources are tight, incentives limited and progress slow. The governance entrepreneur clearly needs to be determined and motivated.

The need for a board and a secretariat

In 2002, Aron Cramer at Business for Social Responsibility (BSR) and Jonas floated a proposal to run a secretarial function for the Voluntary Principles for Human Rights and Security (VPs). A tiny shoestring operation run by Jonas was allowed, without much political or financial commitment from the parties. Perhaps unsurprisingly, few among the companies or governments at that time saw the benefit in investing heavily in this kind of innovative governance. A couple of years later, one of the senior executives who had long been involved with the Voluntary Principles admitted that the companies had made a mistake in failing to realize that it was in their interest to have a dedicated secretariat to provide the initiative with co-ordinating energy and an ultimately stronger mandate. The companies had not been under pressure to give the VPs greater strength and therefore resisted providing the necessary resources and organizational capacity.

Establishing an independent EITI secretariat in Oslo did not require such a leap of faith for the stakeholders. It is important to remember that the EITI only became a formal body with a board and independent secretariat five years after the initiative was launched. Awareness of the challenges of governing natural resources was becoming stronger among companies and the development community. The nascent EITI process had already proven its value by building trust and establishing a clear product—data, national commissions and a standard. By 2007, there was a clear demand for a formal institution.

The EITI International Secretariat began with seven staff—a head, four regional directors, a communications manager and an office manager.[1] The staff has expended gradually—at the time of writing there were still less than 20 full-time staff. The secretariat has avoided the temptations of rapid growth despite the financial means to do so. It has preferred a limited role and flat management structure, still centred around the regional directors/country managers, although they have more support than in the early days, as the role of the EITI has expanded. The International Secretariat still relies heavily on the World Bank and others to provide much of the technical assistance, and all of the financial assistance, to countries implementing and planning to implement the standard. This has helped the organization remain lean, but it also means that many of the key functions on which the success of the EITI are based are provided by outside organizations with varied levels of knowledge, different recruitment practices, conflicting priorities and incentives, and different lines of accountability.

It should also be noted that several interns have also made heroic contributions, many of whom have subsequently become absorbed into permanent roles in the secretariat.[2]

Naturally, funding often soon becomes a contentious topic. While seed funding is commonly available at the beginning of a process, it becomes

---

1   Sam Bartlett, Tim Bittiger, Ingunn Dahle, Leah Krogsund, Francisco Paris and the authors.

2   Six former interns are particularly worthy of mention. Anders Krakenes has been with the secretariat from the beginning and headed the communications team for almost six years before moving on to special projects. Dyveke Rogan was responsible for much of the writing of the 2013 standard and subsequent guidance, as well as taking on a lot of country responsibilities. She joined the senior management team in 2013. Bady Balde has worked on much of the EITI data analysis before leading on Central and North Africa for the secretariat, and becoming a regional director in 2014. Pablo Valverde was employed as an intern by the Norwegian Ministry of Foreign Affairs to set up the International Secretariat in 2007. He became the conference manager for the 2009 Doha conference before going off to work for the Norwegian Oil Fund's Ethics Committee for five years. In 2014, he returned to the secretariat where he heads up country support to the Middle East and North Africa. Shemshat Kasimov came in as intern to support work in Russian and has gone on to support our country work across Asia. Lyydia Kilpi was employed as an intern in 2013, before leading on Lusophone African countries and then on the 2016 global conference. They, and others, demonstrate the secretariat's success at home-growing talent. External recruitment has been more challenging, although the recruitment of six new full-time staff in 2014 has proven successful.

more difficult to find continued sources of revenue. Many organizations spend vast proportions of their time on fundraising. Not only is this distracting, it can also be distorting to the organization's objectives. These are normal challenges for most not-for-profit organizations. It again shows the importance of building networks and tailoring messages.

However, the further challenge for collective governance bodies is how to balance financial support. If a collective governance body is completely funded by one constituency or another, it is vulnerable to accusations of bias and is indeed subject to unbalanced influence. A process for addressing this must be established early on in setting up the independent institution. It is worth nothing that perceptions of the influence of funding vary significantly in different parts of the world. In Northern Europe, for example, it is common that not-for-profit organizations are largely funded

The EITI's funding formula

EITI supporters fund according to the following formula:

- The private sector and supporting countries share principal responsibility for the international management costs of the EITI, with support from civil society organizations and the host government, Norway. The private sector and supporting countries should pay the same.

- The EITI Board ensures that no single constituency or single stakeholder dominates the level of funding.

The number of supporting organizations has grown every year. The 2015 budget for the international management of the EITI was just over US$5 million, which covered salaries and associated costs for about 20 staff, board meetings, travel, consultants and communications. The EITI international management also received considerable support for specific projects, such as funding of ad hoc meetings, training seminars, and so on.

Until 2013, funding of validation had been the responsibility of implementing countries. With the adoption of the new EITI Standard, the EITI international management became responsible for procuring and paying for validation, and later budgets will reflect this.

In-country efforts are funded from a wide range of sources, including the implementing country governments themselves, the World Bank-administered Multi-Donor Trust Fund, and other multilateral and bilateral development agencies.

All of the EITI International Secretariat's income, including off-budget support, expenditure and audited accounts, may be found at https://eiti.org/about/funding.

by governments, without it often being seen as undermining these organizations' ability to criticize funding governments. In North America, on the other hand, funding is many times a more contentious issue, with many not-for-profits having strict funding rules in order to ensure impartiality.

After a board and secretariat have been established and funding secured, comes the discussion about what kind of legal status the organization should have. There are very few models to guide the timing or form of a legal set-up for collective governance.

The EITI's legal entity

A difficult issue to solve for the EITI, and any collaborative governance effort, is to agree its own governance structure. That some kind of organization or secretariat answers to a board or council is straightforward. How the board or council is appointed and to what it is answerable is less obvious. Private companies have shareholders and annual general meetings to which their board reports. Many jurisdictions allow a relatively high degree of flexibility regarding the design of not-for-profit foundations, with the board often appointing itself. For a coalition such as the EITI, it was necessary to devise some kind of body to which the board would report and which would give it adequate legitimacy and accountability.

The stakeholders behind the EITI had to decide early on whether they wanted to establish a treaty-bound intergovernmental body, or a not-for-profit entity that was not recognized by international public law. Turning the EITI into an intergovernmental body would have given it some greater official recognition and standing. It also appealed to some early staff members, as they would have been tax exempt in a country with one of the world's highest taxes. However, it was quickly decided that having a not-for-profit organization was vastly more attractive. It would not take many years of multilateral negotiations, and it avoided even more complicated decision-making involving both state and non-state actors.

In developing this legal entity arrangement, those involved took considerable inspiration from other collaborative efforts, in particular Swiss-based organizations such as the Global Fund to fight AIDS, Tuberculosis and Malaria, Gavi and the Global Alliance for Improved Nutrition.

The EITI's legal status was not established until 2009—six years after the first conference. At the third global conference in Oslo, in 2006, it was decided to form a board and an independent secretariat which would be based in Oslo. It was decided that all EITI stakeholders would be invited to be members of the association. These members would meet once every two years, receive a report from the board, and appoint a new board.

For the first two years, the board and secretariat were essentially parts of an interim not-for-profit members' association with two members: its chair, Peter Eigen, and Ambassador Tormod Cappelen Endresen of the Norwegian Ministry of Foreign Affairs.

In 2009, the interim members' association was replaced with a more permanent not-for-profit members' association under Norwegian law. It has a board with representatives from governments, civil society and companies which is answerable to a conference and members' meeting, which is convened once every two years. The period between these conferences effectively forms the term of office for board members, including the chair. These arrangements are set out in the EITI's Articles of Association, which were slightly amended in 2011 and 2013, including the hosting of the members' meeting only every three years from 2013.[3]

It was complicated to devise the rules for the members, as the EITI was supported by stakeholders with different interests. The members were therefore divided into three constituencies: governments, companies and civil society. To safeguard the organization from one constituency signing up large number of members that might be mobilized at a members' meeting to vote through a board of their liking, the votes were weighted giving each constituency the same influence.[4]

The last aspect of the governance architecture related to the forms and rules for the constituencies to decide on how to nominate members to the board. Even if it was going to be up to them to agree how they wished to be represented on the board, the organization had to develop some safeguards for ensuring that good processes were followed by the various constituencies. The EITI therefore adopted a set of constituency guidelines including, for example, rules about how the constituencies must have transparent procedures for agreeing board nominations.

By the way, a modern quirk also prevented the EITI from becoming a Norwegian foundation. As with Norwegian private companies, a Norwegian foundation must have a minimum of 40% female board membership, and the stakeholders behind the EITI could not be guaranteed to nominate that many women to the board. However, this quota is not a requirement of a members' association.

3   The Articles of Association set out the provisions for how the EITI Association is governed (EITI, 2013c). A "member" of the EITI Association is a personal representative of a country (meaning state), company, organization or legal entity that is appointed by a constituency as set out in Articles 5.2 and 5.3.

4   Each constituency decides its own rules governing the appointment of members of the EITI Association. The constituencies may have unequal numbers of members (for example, the countries have one representative per implementing country and one per supporting country, while the companies have one representative per supporting company and institutional investor), but weighting ensures that each constituency has the same number of actual votes.

7.2 Accountability: build capabilities *alongside* accountability

A particular challenge for a multi-stakeholder governance effort is to have governance arrangements that provide adequate oversight of the initiative itself, while at the same time acknowledging that the interests of various stakeholders' are reflected. If a person serves on the board of a for-profit corporation, that person has fiduciary and other responsibilities towards the corporation. Those obligations will take precedence over the demands of ownership that the board member may represent. The case of the EITI is more complex. As we have seen, all board members except the chair are representatives of their respective constituencies. There can be tensions between the EITI's interests and those of, for example, some of the supporting companies.

The EITI and codes of conduct

As in any loose and open movement, the lines of accountability within the various EITI bodies are necessarily somewhat messy. The EITI Board is accountable to its members, but this link is weak. The secretariat is accountable to the board through its head, whom the board appoints.[5] The implementing countries each have EITI governance structures with no accountability to the international structures (apart from being found compliant, candidate, suspended or delisted from the international system). Yet these national bodies and the individual members use and benefit from the EITI branding.

In formal constitutional terms, there is clearly an accountability deficit for EITI office holders, i.e. anyone who serves on the board, secretariat, or national commissions and secretariat. A scandal on funding, fraud, opacity, per diems or any minor issue could blow the whole EITI off course. The risk was of course inherent in that the EITI was created to address exactly this kind of accountability deficit in the first place. The EITI could not completely safeguard against these risks, but it could build a mechanism for bolstering accountability: deterrents and processes for appropriate swift action if faced with these challenges.

In light of this, in 2014 the EITI belatedly established a code of conduct for EITI office holders.[6] The code outlines the responsibilities of, or proper

5   According to Article 17.1, the head of the secretariat is responsible for the activities of the secretariat and reports to the EITI Board through the chair.

6   EITI (2014) also sets out the association's policy on openness, outlining the exceptional cases in which EITI papers and correspondence might be confidential.

practices for, individual EITI office holders and EITI institutions at the international or national level, including their decisions, their procedures and their systems. It draws on cases in which a lack of a code had made action against perceived misconduct by individual post-holders difficult. The EITI has a governance committee to recommend to the board any appropriate action related to individual office holders for breaching the code.

The code of conduct contains the provision that "EITI Office Holders should dedicate themselves to be leading by example and should represent the interests and mission of the EITI in good faith and with honesty, integrity, due diligence and reasonable competence in a manner that preserves and enhances public confidence in their integrity and the integrity of the EITI, and ensuring that his or her association with the EITI remains in good standing at all times". It contains further provisions on respect for others; professionalism; anti-discrimination; confidentiality; expenditure of EITI resources and use of EITI property; conflict of interest and abuse of position; gifts, trips and entertainment; and mechanisms for implementation and breaches of the code.

Of course this does not insure against abuse, but it does provide a mechanism for addressing it.

The EITI and per diems

All development practitioners know about the distortions created by per diems, whether they are provided by international bodies, national bodies or are absent. It is, of course, notoriously difficult to legislate for "appropriate" compensation/reimbursement for attending meetings and the authors do not offer new insights. Having representatives from across a country as large as Indonesia attend a meeting in Jakarta clearly requires more compensation than across Sao Tome. However, there have obviously been cases of abuse of per diems by national commissions. Some have offered flat-rate attendance to members of over US$1,000 for attending each meeting. This creates an enormous incentive for holding more meetings, distorts and delays decision-making and skews representation. The EITI has no rules against this but encourages such matters to be referred to the board and national commissions are asked to account for their policies.

The EITI International Secretariat does not offer per diems for its own staff—just reimbursements. It provides travel and modest per diems for board members from Southern civil society. It also provides a communications bursary for board members from Southern civil society to enable them to participate fully in EITI committees, constituency meetings, to print out papers and so on (this is managed by Publish What You Pay International).

7.3 Demand: ensure that the institutions are and continue to be demand-driven

This book has already covered the need for conflict to bring parties to the table and for sustained conflict to keep them there. It has also noted the need for a "product"—something for the collective governance group to do.

A standard is one such product and therefore makes for a fruitful area for collective governance. However, a standard has to be robust, tough and meaningful, while its implementation must be relevant, appropriate and flexible. This inherent tension—between ownership and relevance by implementers and central robustness by the "governors" of the collective governance—is healthy but is tough to steer. Implementers complain that the standard is not flexible enough to accommodate their circumstances, while central stakeholders complain that the brand is in danger of being damaged by diverse implementation. Conceptually, the **process** should be owned and demanded by the implementers, but the **standard** itself should belong to the central body, on which the implementers are represented. The experience of the EITI is that, in practice, the lines are not easily drawn.

The EITI: a global standard for local transparency

It's a good tag line, but it is an inherent tension. In principle, the tension should be managed by the fact that the countries themselves partly own the global standard through their board representative. If implementing country X is not happy about the standard, it should use its representative on the board to change the standard. In reality this rarely happens, and there are plenty of accusations from countries that "Oslo" (where the EITI International Secretariat is based) is dictating to them and undermining national sovereignty, even though it is the board that makes decisions on their status and it is the country's sovereign decision whether to implement the standard or not.

There are occasionally threats that countries will do their own EITI outside the international process if they feel hard done by. Such competing EITI processes could potentially be a challenge to the credibility of the international EITI—it rises and falls on the basis of being a global standard. Ethiopia and Zimbabwe have both at various times implemented parallel national EITI processes—though neither abandoned hope of joining the international club in due course.

This could, of course, all be shrugged off as sour grapes, except that the aggrieved countries often have a valid point. They might be using the EITI process for all sorts of exciting, important and meaningful discussion, but are being prevented from compliance by some minor technicality that has no real world relevance. Often they might have trapped themselves by committing to cover too many negligible payments in the report, only to find themselves having to chase tiny figures in order to meet the standard.

The EITI clearly has much more to do on promoting collective ownership of the standard and promoting awareness of the technical pitfalls.

Standards have the other chief benefit of containing an inherent reward mechanism. In the case of the EITI, this is the assessment of the standard: the board's role in bestowing compliance, candidacy, suspension or delisting on implementing countries. As noted previously, while far from perfect for assessing performance, it has been essential in sustaining demand. Like other aspects of the governance of collective governance, these reward mechanisms should be allowed to evolve and become more sophisticated, or complemented, degraded or even replaced, as the consensus moves.

Lessons from chapter

- Build modest institutions with clear mandates and which are fit for purpose. Adapt the mandate, the institutions, the staffing and the funding, incrementally.
- Establish codes of conduct for stakeholders and staff.
- Build in a reward mechanism (such as compliance), though make sure that it is not seen as an end-point.
- Make sure that implementers feel that they have ownership of the process and do not want to walk away.

8
Saying goodbye: sunset clauses and appreciating the life-cycle of institutions

Without good design or critique, most institutions eventually exist at some level only to justify their own existence and continuity. Institutions should never, of course, become ends in themselves. As stated before, collective governance should exist only to strengthen governments to do the job by themselves. Only governance by governments leads to sustained societal change. Yet it is notoriously hard for any initiatives, collective governance or otherwise, to put themselves out of business. "Sustainability" often becomes the code for financial self-justification rather than creating a permanent solution. Despite its mesmerizing attraction, empire building must be resisted if a body is going to stay fit, relevant and nimble. Sunset clauses, even if well intentioned, get forgotten or permanently postponed. Most institutions work best but achieve least within a silo. They are not incentivized to link with other bodies working on complementary or overlapping mandates.

The dynamics of the for-profit sector that often lead to mergers and acquisitions do not exist with not-for-profits. Organizations may identify savings by merging with other similar organizations but, unfortunately, there is often a fear that funding will be harder to raise for one organization than it was for what was previously two organizations. The management may also have less self-interest in merging not-for-profit activities than the management of profit-making companies. Superfluous managers in profit-making companies can relatively easily be financially compensated for loss of a leadership job as a result of a company takeover or merger. For not-for-profits, it is often not possible to compensate in an attractive way and the market for managers is less liquid, making managers less interested in contributing to a merger or takeover that may result in them losing their job.

Even when bodies, such as collective governance institutions, are established essentially to bridge a governance gap, they rarely seek to be absorbed into mainstream government processes.

A sunset clause was, at least for the EITI, probably a condition for corporate support for the process. Companies that were deciding whether or not to support the EITI wanted to be sure that they were not making an open-ended commitment. Their support was dependent on assurances that they could end their support if they so wanted.

The key institution, the collective governance body or the board, is responsible for making the process successful, relevant and meaningful, but not only are board members the worst judge of when to wrap up the process, they can even act as a bulwark against mainstreaming the process. They can end up knowingly or unknowingly withholding information that informs public debate. They can block others from joining in the key public debate forums. The brand, the standard, the reports, the conferences, as well as the institutions, can all prevent the positive drift towards integration into strengthened government systems.

With all internal processes inherently favouring the permanency of institution, collective governance initiatives need a systematic process to move from "consensus shifting" to "adaptation/graduation" through to "institution closing". There needs to be an alarm system to signal this evolution and particularly when to break out of the institutions. This should be an evaluation for independently assessing the continued usefulness of the institutions—are they still relevant, have they been mainstreamed or have they failed? Ideally, this regular self-assessment and/or "sunset clause" should be written into the legal entity.

When the collective governance issue has begun its inevitable drift into government systems, a collective governance body has two choices:

1. To dismantle itself

2. To reinvent itself

For the EITI, it is difficult to know if the authors are prey to the self-justification mentioned above, but it appears that there is still a long way to go before the issue of natural resource management can be addressed by government alone. What is clear is that the EITI cannot be said to have completed the ambitions set out in its principles. Just because the job is not fully done, however, is not a reason to resist reinvention. As noted before, simply being a body to assess compliance when two-thirds have already passed that milestone does not look like a sustainable model. Revising the standard has already gone a long way in providing new meaning to the process and the institutions. But there will be a need to consider new functions. These might include more narrative assessments and expert evaluations of countries' management of their extractive sector. Alternatively or additionally, it might involve increasingly becoming a forum for dialogue about company best practice on tricky due diligence issues that might be informed by more data in the public domain, such as how to know the history of assets that you are buying.

The EITI and sunset clauses

The EITI International Secretariat was established in 2007 with a three-year sunset clause, i.e. with the intention being that it should close after three years. This was written into the original process as an insurance against failure. In 2009, the sunset clause was changed and the fixed closure date was replaced with a provision stating that the secretariat should remain open until it is no longer required.

At the time of writing, the EITI continues to grow and to deepen, and there remains a clear role for a global owner (the board) and custodian (the International Secretariat) of the EITI Standard. However, as more and more countries pass the standard and develop their varied EITI models, hopefully involving more integration into their governmental systems, the question of what the purpose the global EITI Board serves will become more pertinent. In 2015, only two board meetings were planned instead of the three or four that had taken place in the previous seven years. To stay relevant, the board will need to consider whether to develop some sort of scoring or more sophisticated system of narrative assessment. It might also consider a graduating system beyond the minimum standard. The board will increasingly need to consider whether a country that fulfils the principles of the EITI by producing reliable, timely and comprehensive government and company data across the whole governance chain, should be excused or "graduate" from producing a specific EITI report. In Norway, for example, five years of reports have been produced that showed no discrepancy between government and company information, all of which is available elsewhere. If the EITI process provided an easy-to-use link or portal to this information as well as a forum for informing public debate, should the board still require a mechanical process of producing a separate report or could the country somehow graduate from this requirement?

Without a board, there will be no role for the secretariat, although that body could adapt to become a resource centre for good practice and guidance drawn from the processes around the world.

These are the signs that the EITI international institutions have achieved their final purpose and thus have come to the end of their useful life:

- Reliable information required and encouraged in the EITI Standard is increasingly produced within government systems.

- The EITI reports in each country become short additional commentaries—diverse in nature and not requiring significant central guidance.

- The national commissions and secretariats are confident of their own role and are exchanging information and experience among themselves.

- The international database of extractive sector information has become integrated into IMF, World Bank or United Nations systems through good coding in EITI reports and better data capture by the international system.

- Other bodies, such as universities and international agencies, take on the role of resource centres for good practice in the governance of the extractives sector.

What is clear in this analysis is that the authors are not suggesting that the end of the EITI international institutions should signal the end of the EITI processes in the implementing countries. That is, of course, an issue for the countries themselves and will be determined by whether there remains a conflict or information gap that cannot be resolved by government alone.

Lessons from chapter

- Avoid institutional inertia through regular external review of the institution, as well as of the collective governance model in general.

- When the drift towards better governance by governments becomes inevitable, either dismantle or reinvent.

9
How to be a governance entrepreneur: checklist

Part 3 of this book has sought to guide a governance entrepreneur through the challenges of managing a collective governance process. The authors clearly have experience with one particular collective action initiative which has grown and evolved somewhat organically and somewhat in a directed fashion. While this does not give them great authority to prescribe a route for other efforts, what follows is the authors' best advice on the broad direction of travel, rather than a prescriptive step-by-step guide.

1. Bring the parties to the table.

2. Agree a point of consensus.

3. Implement a pilot.

4. Move the consensus—and consolidate around an agreed product.

5. Establish institutions with political, financial and legal under pinning.

6. Continue to move the consensus—through evaluation, review, etc.

7. Incentivize integration/graduation of processes into government systems.

8. Close the institutions.

Drawing on the previous chapters, the above steps involve the following lessons:

Make sure that the preconditions for collective governance exist

The collective governance route is likely to work best when the following conditions exist:

- Clear government failure where there is a lack of trust in both government and companies.
- A problem that requires long-term co-operation, not a one-off deal.
- All players have something to gain and a lot to lose.
- Enduring conflict to continue to get the players around the table and to keep them there.
- A manageable number of main companies.

In short, collective governance is an expression of crisis.

Build trust *through* momentum

- Successful collective governance depends on a sophisticated consensus.
- Collective governance has to look beyond voluntary vs. mandatory battle lines, and instead focus on shorter term actions leading also to trust building.
- Don't get bogged down by discussions about representation, profile and mandate. Get over it—the right people will come if the right issues are being discussed.
- On the other hand, accept some focus on process rather than substance, especially from the weak who are unfamiliar with the topics of substance, and from the strong who might be uncomfortable with the content of the substance.
- Build capacity and address power imbalance by building substance.
- Attention, network building, capacity building and facilitation are required for stakeholder groups to ensure that they can contribute appropriately.
- Allow the standard to be adaptable and attractive to all potential implementers.
- Make sure that the intent of the rules outweighs the letter of the rules, so as to encourage progress while being tough where political will is missing.
- Be conscious of giving each stakeholder some sense of a win in critical debates.

Move the consensus from the narrow to the meaningful

- Focus on the possible, not the meaningful. The meaningful can come later.
- Hold your nerve against criticism of irrelevant focus, but make sure that the process is linked to other debates and processes.
- Keep challenging the stakeholders to move the consensus.
- Ensure that the rules are robust but do not lock in inflexibility.
- Build in systems to refresh the process, such as evaluation, strategy review, etc.

Get the most from people

- Recognize your role as a spider in a web. Build a support network by meeting face to face with potential friends of the process—find out their motivations and tailor your messages.
- In the absence of results, get endorsements and explain the process.
- Look for opportunities and make sure that you have enough friends in the right places to take advantage of them.
- Use support networks to make best use of the product to generate debate.
- Don't oversell the process. Sell the product.
- Help support networks to use the data, turn it into stories and stimulate analysis and debate.
- Seek out leadership skills at all levels.
- Embrace and acknowledge the heroism of compromise through documentation, speeches and discussion.

Govern the governance

- Build modest institutions with clear mandates and which are fit for purpose. Adapt the mandate, the institutions, the staffing and the funding, incrementally.
- Establish codes of conduct for stakeholders and staff.
- Build in a reward mechanism (such as compliance), though make sure that it is not seen as an end-point.
- Make sure that implementers feel that they have ownership of the process and do not want to walk away.

Be able to say goodbye

- Avoid institutional inertia through regular external review of the institution, as well as of the collective governance model in general.
- When the drift towards better governance by governments becomes inevitable, either dismantle or reinvent.

Part 4:
Recommendations
and conclusions

10
Recommendations

As the previous chapters have demonstrated, collective governance is difficult. For them to work, they will need significant support from the international community—governments, development agencies, civil society and companies. The governance entrepreneur will need extensive support. As we have seen throughout this book, different stakeholders have different roles and contributions they can make to support collective governance efforts. They also encounter challenges that are often particular to their constituency. Development agencies often speak in glowing terms about collective governance. Still, many of their procedures and frameworks stifle it. Similarly, companies and civil society praise such collaboration while sticking to conventional ways of working. The authors therefore make ten recommendations for embracing and incentivizing effective collective governance: one for domestic governments, four for the international diplomatic and development community, two for civil society, two for companies and one for all actors.

Recommendation 1: Domestic government to convene and participate

The title of this book, *Beyond Governments*, should not be understood as a call for governments to avoid engaging and taking responsibility. Nor should it be interpreted as a call for other actors to take over the responsibilities of governments. The title is merely a recognition that some challenges cannot be effectively addressed by governments on their own; that they need to collaborate innovatively with companies and civil society representatives in order to strengthen themselves. At a minimum, domestic governments should participate in collaborative governance efforts. This may not seem a great ask, but may well require a radically different mind-set from the more common prevailing stance of simply issuing and ensuring the following of laws and regulations. The presence of government representatives can often be enough endorsement to make other actors contribute. An ambitious government can do more than any one other body to convene and assist in bringing other actors together. On the flipside, without government involvement, collective governance cannot prosper. Essentially, discussion about public policy without government in the room is neither "collective" nor "governance".

Governments: be bold, reform and participate.

Recommendation 2: Diplomats and development agencies to bring a peace negotiation mind-set to their support for collective governance

In times when development agencies increasingly have to demonstrate impact and value for money, many have seen their administrative resources shrink in relation to the development assistance disbursements. An increasing scepticism in many OECD countries about the effectiveness of aid has put many agencies in a defensive mind-set. Being able to demonstrate specific tangible outcomes, such as bed nets or textbooks, becomes irresistible.

The authors are strong supporters of such aid, and of the need for sound performance indicators, yet supporting good governance and, in particular, collective governance, does not work like that. It needs a different mind-set from "$X buys you Y number of lives saved" or "$A stimulates B% increase in GDP". As shown in Chapter 5, the outcomes of collective governance are uncertain and do not follow a linear pattern of the type foreseen in logical frameworks or theory of change analysis. Furthermore, the grants provided are relatively small compared with the support and engagement required.

For diplomats and international development workers, collective governance is a fundamentally different form of development support: the inputs are staff-intensive, and the outcomes are unpredictable, intangible, non-attributable and long term. Yet, like peace or trade negotiations, the potential benefits are enormous. In the face of resistance and inertia, the agencies need to make the case for wandering into the dark.

Like companies, development agencies engaged in collective governance need to develop a different business model. They need to see themselves as more than funders or technicians. The international development community should assess the enduring governance challenges at the global, national and subnational level. Then they should accept the lack of power that any single institution or body has to manage the complex multi-dimensional forces that are required to address the specific governance challenge. Then they should study whether the preconditions exist for collective governance. If they do, they should recognize their almost unique position as conveners, incubators and facilitators of such models at the international level. They could establish model pilot programmes, peer reviews, expert panels, governance indices, codes and standards, etc., to incentivize and support such collective governance efforts with due recognition of their evolving and time-consuming nature.

Most radically, they will have to put aside their beloved monitoring and evaluation tools and theory of change approaches to be more responsive to adaptive models of development and governance entrepreneurship. But they should not do this in a reckless manner. There are many vague initiatives only too willing to have funders throw out the conventional monitoring and evaluation requirements in favour of woolly self-justification, such as "trust building" and other such concepts that have led to 76% partnership failure.[1] No—this should only be done on the basis of a robust analysis of the preconditions for successful collective governance as set out in Chapter 3. Is there the fuel of conflict ready to fire this collective governance process?

The EITI is a good example of how the development agencies gambled on a different model. As explained before, for the EITI there were no goals and logical frameworks, no agreed objectives, no theories of change. The UK and German governments and the World Bank have been particularly active in bringing together other stakeholders and lifting the agenda to the global stage.

There were many occasions in which the lack of these conventional development frameworks threatened to scupper the chances of international support and funding for the EITI. It became difficult for some development agencies to justify funding to a programme that did not appear to have a clear objective or end-point. The EITI model could easily have been

---

1   Biermann *et al.*, 2012.

adapted to have a clearer theory of change, with a straightforward objective, indicators and milestones. However, such an approach would either have been trite or lost the confidence of most of the stakeholders. This story tells us a lot about the nature of collective governance as an adaptive, not planned, form of programming.

Recommendation 3: Development agencies to support civil society's efforts to contribute to public policy

The state will continue to be the main body for public policy decision-making and, to a lesser extent, delivery. Building capacity in the public sector to undertake that role has rightly been the focus of development agencies. Being mostly public-sector bodies themselves, international development agencies have tended to be most comfortable building the capacity of other public-sector bodies. With a burgeoning global middle class and unprecedented access to data and social media, the capacity to make good public policy decisions will not, in the future, have to reside so exclusively with the public sector. Instead, the challenge will be to harness the forces of technology, media and education to inform a generation of public-spirited citizens, and to provide them with a forum in which to debate and shape public policy.

While there has been enormous support for civil society activism and capacity building from international development agencies for many decades, it might be useful to explore more deeply the models for global coalition building around specific campaigns, backed up with efforts to create expertise and use of data, on which the EITI rode.

The Publish What You Pay (PWYP) coalition provides a useful template. The efforts of PWYP have led to significant improvements in the policies and practices of governments and companies in the extractive sector through the building of civil society rather than government. The coalition's broad organizational structure, clear campaigning objective and lean bureaucracy have enabled the campaign to be broadly informed by activists in the affected countries and for those activists to learn from other another.[2] Furthermore, the establishment of national PWYP coalitions facilitates more direct support of local civil society. The World Bank has been providing

2   Van Oranje and Parham, 2009.

direct funding to national civil society engaged in the EITI process across five countries[3] and further support to another five through the Natural Resource Governance Institute.[4]

In addition, the capacity building, research and advocacy efforts of the Natural Resource Governance Institute were needed not only to provide rigorous expert support to the civil society groups across the world, but also to bring together and use the data arising from the process.

Recommendation 4: International development bodies to redefine official development assistance, to improve the accommodation of global standards

Developing countries do not necessarily consider themselves to be a cohort—that is a developed country ascription. The mind-set of development agencies will need to shift away from that attitude. Good governance in developing countries is increasingly influenced by the systems and standards practised by all countries and which might need support from development agencies. There is a need for international development bodies such as the OECD to redefine development assistance, to improve the accommodation of global standards.

While global collective governance processes such as the Open Government Partnership are good for development and no doubt contributed to catalysing EITI progress in countries where resource governance is critical (including the Philippines, Mexico, Ukraine, Colombia and Myanmar), they create a problem for development partner funding. In order to qualify for Official Development Assistance (ODA), all aid money must benefit "developing countries".[5] Yet a global standard is, by definition, for all countries. The stronger the global standard, the stronger the incentive

3   World Bank Development Grant Facility, 2007.
4   Natural Resource Governance Institute, 2015.
5   Official Development Assistance (ODA): "Flows of official financing administered with the promotion of the economic development and welfare of developing countries as the main objective, and which are concessional in character … By convention, ODA flows comprise contributions of donor government agencies, at all levels, to developing countries ('bilateral ODA') and to multilateral institutions" (OECD Glossary of Statistical Terms, http://stats.oecd.org/glossary/detail.asp?ID=6043).

for poor countries, and thus the benefit for citizens of those countries. The World Bank-administered Multi-Donor Trust Fund for EITI technical assistance still cannot support EITI implementation in OECD countries. Support for the EITI International Secretariat failed ODA qualification for several years before 2010. It is clear that, as codes and standards and development efforts such as the Open Government Partnership become more mainstream, the rules on ODA will need to be revisited.

Recommendation 5: Diplomats and development agencies to build on models that help countries through the instability between autocracy and open democracy

Hillary Clinton, then US Secretary of State, characterized the development debate as having moved from an issue of Northern and Southern countries to one of open and closed countries. She was right, and the development agencies need to be more focused on how to help countries make the difficult transition. Myanmar and several countries in the Middle East and North Africa are presently experiencing upheavals in the process. The desire to stand up to old autocrats appears to be spreading across Africa with the fall of Compaore in Burkina Faso as the latest. There are few development models that are responsive enough to complex power shifts and factional upheaval. Again, the authors would not wish to oversell collective governance, but its role in bringing the right stakeholders around a table, while more formal institutions of democracy are being constructed, could be of use. The government of Myanmar has committed itself with full political backing to the EITI process. It is being implemented in Iraq and many countries in Africa which are emerging from closed policy environments. The EITI journey for most of these has not been straightforward, but each, with continued good luck, appears to be on the upward slope of Bremmer's J curve.

Recommendation 6: Civil society organizations: engage and inform debate as well as campaign

Campaign organizations increasingly recognize two realities. One, that they naturally seek wins. Two, that business can lift a lot more people out of poverty than they can.

On the first, everyone knows that governance issues are more complex than simple wins or slogans such as "Drop the Debt", "Make Poverty History" or even "Publish What You Pay". Everyone also knows that they are highly effective for raising awareness and mobilizing funding and support. Campaigning is particularly effective when there is a clear and known target, such as a large energy company, even if lesser-known companies often have further to go in terms of operating to high standards. Short, snappy slogans become even more important in a world of Twitter. Campaigning for legislation, such as a minimum wage, gives a clear moment of victory and is relatively easy to raise financial support for. Less time is spent on assessing whether the legislation has achieved the intended goal, i.e. reduced poverty. The legislation might not be enforced, or might create perverse incentives such as putting workers out of business, or it might simply increase the informal economy. While such campaigning is clearly important, it is also necessary that campaign groups remain engaged beyond the initial "win" to see the policy process to its desired conclusion.

The United States and European Union have passed legislation requiring extractive companies to publish what they pay to governments around the world. Now it is necessary for international civil society to remain engaged to ensure that the data is reliable, used, analysed and compared, and informs debate in the countries in which the payments are made. For this to happen requires a change in mind-set for most civil society organizations from simply asking for more data because they do not trust the other actors, to using that data to generate real accountability. In this aspect, the work of the Natural Resource Governance Institute is admirable.

On the second, very few civil society organizations still consider business to be the enemy of poverty reduction. Most of them seek to constructively engage with big business—which was not so common 15 or 20 years ago. This requires some restructuring of how many civil society organizations operate. They need to equip themselves both for advocacy and for engagement. They might also need to deepen their analysis and capacity-building capabilities. Having civil society activists engaging more in long-term public policy debates with governments and with companies is complex, time-consuming and perhaps problematic for fundraising, but will be necessary for a changing political landscape.

Recommendation 7: International civil society to bring people together and enhance knowledge sharing, but avoid dominating the citizen voice

International civil society is the fire to keep a collective governance process going and especially to keep companies at the table. The EITI would not have happened without Publish What You Pay. But with that comes a lot of responsibility—to build coalitions across countries, to guide and support national civil society efforts and to use the information from the process. The closer the national civil society is connected with the global coalition and the stronger that global coalition is, the faster their mutual learning curve.[6] Coalitions are about not just about joint learning, they are about the power of acting in numbers.

However, for the international coalitions to be most effective they need to respond more to the interests of their national stakeholders than their international funders. This requires that they have strong governance mechanisms which give sufficient voice to the affected citizens. It also requires them to assess whether they can be both advocate and co-ordinator for such diverse interests.

Recommendation 8: Companies to develop more dynamic business models that reflect a desire for justice rather than just charity

Companies need to develop more dynamic business models, with better rewards for leadership, that benefit both the company and society in the longer term. In many ways, it is companies that need to undertake the bravest and most radical change in approach. They need to think outside the conventional business model by incentivizing and rewarding staff who engage in collective governance efforts which promote their social licence

6   Van Oranje and Parham, 2009.

to operate, level the playing field with other companies and promote public understanding of their business. They need to stop thinking about supporting these processes as charity, and more about how all their business practices promote a just society. This means that engaging in collective governance must belong to the mainstream of their business—it must be closely linked into their investment policy.

This is no truer than when fighting corruption. The combating of corruption starts with following laws and regulations, but that on its own is not enough. As seen with the EITI, corruption often suffers from a collective action dilemma, or Prisoner's Dilemma, meaning that it would be against any one individual's self-interest to undertake alone the action that would lead to everyone's longer-term benefit.

Companies are daily invited to sponsor, endorse and participate in conferences, dialogues and processes aimed at improving the state of the world. The benefits of all such engagements are hard to measure, quantify and see on the bottom line. The authors of this book do not expect companies to respond positively to all of these invitations. Many of them are costly distractions. The difficulty of distinguishing the wheat from the chaff should not, however, be an excuse for not engaging more widely. The difficulties in measuring the financial results of longer-term and wider engagement make it necessary for companies to apply a strategic mind-set, a ruthless prioritization, to involvement. To do so, companies need to be policy savvy, distinguishing between talk leading to little change, and actions.

It is of course difficult to identify which processes will deliver concrete actions and results, and it can seem too complicated to try, but that is not an excuse to do nothing: serious company engagement is often required for good outcomes. In the energy and mining sectors we have seen examples of leadership, such as:

- Being transparent themselves. For example, Rio Tinto and Statoil have been publishing country-by-country payments in a single report for several years now. So has Shell, though it has not published for all its countries of operation. Tullow Oil in 2014 became the first company to publish its payments on a project-by-project basis, pre-empting the exact wording of the EU Accounting Directive and the US Dodd-Frank 1504 legislative requirements.

- Telling host governments that high standards matter to them. High standards do not threaten their relationships.

- Promoting public understanding of their sector. For example, BP, Chevron, Shell and Total have provided training to citizens in Myanmar about how oil contracts work.

- Integrating "public policy-making" of the type promoted in collective governance into all their work, and incentivizing country managers to consider spending time in collective governance groups developing public policy as core business.

The authors are no experts on cost–benefit analyses and it is difficult not to sound trite on these matters. Having said that, they expect that there could be an interesting freakonomic analysis to be written on the cost of operating an oil rig compared to participation in a national multi-stakeholder group like the EITI. The tangible is far more satisfying and apparently productive, but the intangible might be where big money can be made and saved.

Recommendation 9: Companies: act now or you will be mandated

The mandatory disclosure debate in the extractive industries shows that if companies do not engage with governments and civil society at the right time on complex governance issues, they will not have a say in how the policy is made. Instead, laws and regulations will be imposed. "Do not wait for more binding legislation before doing something," concluded the Swiss State Secretary Yves Rossier when addressing a gathering of energy trading companies in 2014.

The authors believe that we are approaching such a stage with hidden ownership. NGO pressure on companies is increasing and will likely lead to change, which hosting governments will reluctantly have to accept. The framework of this book suggests that such an issue would benefit from a collective approach and the EITI is likely to provide a venue where such actors can meet. The trick for companies will be how to balance the demand to provide unilateral leadership to avoid embarrassment, set the agenda and to claim progressiveness, but not fall foul of their host governments.

Collective governance might be time-consuming, frustrating and, at times, seemingly pointless, but the alternatives might be much more harmful to the business or difficult to handle. Companies: act now or you will be mandated to do so.

Recommendation 10: All actors to promote good practice and peer learning

If collaborative governance efforts are to shape and contribute to improved policy, they must learn from each other. Facilitating capacity building and mutual learning is a critical part of making it work. This might include:

- Staff exchanges between corresponding collective governance efforts in different countries.

- Twinning corresponding collective governance efforts in different countries.

- Global and regional hubs of expertise.

- Staff exchanges between the stakeholders, such as government staff working in a company, or company staff working in a civil society organization, etc.

- Development of guidance materials, including the documentation of good practice.

- Facilitating support networks, such as regular newsletters, interactive e-mail or social media sites for exchanges of ideas, etc.

- Seminars, conferences, workshops, etc. on related topics can be instructive and help build co-ordination and linkages.

- Formal training (general or theme-specific) for all stakeholders or for specific stakeholders; on the job or away from the office, etc.

- The development of better reward mechanisms for engagement.

11
Applicability of governance entrepreneurship in other sectors

Part 3 explained the preconditions, skills and institutions for collective governance. The clear lesson is that collective governance is not widely applicable. Nor it is easy. However, the world is a messy place and there are many seemingly intractable policy areas where there are no easy governance models. Multilateral approaches, as seen in the United Nations or the European Union, have proven effective for building trust between states and setting global agendas, but for issues needing fast action they have been protracted, cumbersome and bureaucratic. It might be unsurprising that they have been slow to make progress on the big global governance issues of climate change, trade and poverty; they have somewhat more worryingly been slow or ineffective on military intervention, sanctions and development intervention.

With so few effective options around, hard though it is, there is surely more space for trying out collective governance. According to the authors' preconditions, collective governance can only be successful in areas in which there is a shortfall of reliable information or speculation leading to volatility, and a business case which says, "If not collective governance, then significant investor risk and/or legislation." The commitment to work collectively has to be sincere. Companies in the extractive sector learned, for example, that vague and rhetorical commitments to transparency were not enough to convince activists and legislators that good governance was assured. It resulted in legislation that many of them considered to be against their best interests. Without companies being more alert to the emerging need to work together with civil society, this pattern can be expected to be repeated in other sectors.

Here the authors look at four areas in which the model might be applied:

1. International financial flows
2. Public infrastructure planning
3. International land acquisitions
4. Press regulation

These four areas deliberately cover a diverse range of topics and a cross-section of global governance issues (international financial flows), national governance issues with international dimensions (international land acquisitions), national governance issues (press regulation) and subnational governance issues (public infrastructure planning). The case of international financial flows is the most audacious suggestion and the authors do not suggest that the conditions exist for collective governance in that area yet—only that the international community should be on high alert for the conditions to arise.

The authors claim no expertise in these topics and fully recognize that greater minds than theirs have been pondering these public policy areas for far longer than they have. Nor do they suggest that collective governance is some magic bullet to resolve the challenges. As with the EITI, it can be a necessary and evolving first step. Nonetheless, drawing on the framework for collective governance set out in this book, these propositions are humbly submitted for greater exploration and modest pilot programming.

Whether or not any of them really are fruitful for collective governance, it is clear that there needs to be more effort to share learning across existing collective governance efforts.

11.1 International financial flows

Global Financial Integrity estimated that the cumulative stock of illicit financial flows from Africa amounted to US$865 billion between 1970 and 2008, and that the figure could have been as high as US$1.8 trillion (Tax Justice Network Africa, 2011: 12). Between 2008 and 2010, transfer mispricing cost Africa a further average US$38.4 billion every year.[1] These figures dwarf Africa's inflows from either international aid or foreign direct investment

---

[1]   Africa Progress Panel, 2013: 66.

put together. Not only does this represent a scandalous loss of revenue for Africa, money laundering and companies effectively volunteering how much and where to pay tax is a threat for governance everywhere.

Most international financial flows are facilitated by a few international banks. A company receiving revenue in country A, but paying tax on the profit generated from that revenue in country B, usually transfers the money through an international bank account. Like the oil and gas industry, the costs of entry into the major international banking sector are high and the actors are consequently few in number.

Following the financial collapse, the behaviour of these banks is under unprecedented scrutiny. The recent attention on transfer pricing has heightened the demand. Governments around the world are considering a range of regulatory options. Civil society campaigners are pushing for greater transparency of international flows and public registers of the beneficial owners (i.e. individuals) of companies, trust funds and foundations. The aim is to see whose money is going where and thus hold them to account. How did they make their money? How did they secure that deal? Why aren't they making a profit where they are making the revenue? Etc.

The first challenge for national regulation is that these flows are, by definition, trans-boundary. Even if more transparency was reported at one end, it would say little about where the money was going. Tracking flows would be impossible without multilateral agreement. Unilateral action is difficult because governments fear driving these banks out of their country; banks themselves fear losing customers to banks operating in a less transparent jurisdiction; companies fear loss of control of their tax arrangements which give them a competitive advantage. The present system is a threat to governance and business models, yet unilateral action is economic or commercial suicide—a classic Prisoner's Dilemma with a resulting race to the bottom.

At the G8 Summit at Lough Erne in 2013, the UK prime minister, David Cameron, committed to draw up a UK register of beneficial owners of companies that was open to the UK tax authorities. His stance was supported by the Confederation of British Industry who described it as a "no brainer".[2] It is unsurprising that international companies support beneficial ownership as they see it as a means to a more level playing field when they compete for business with domestic companies who might have closer political ties to the host government.

2   Wintour, 2013.

A register of beneficial ownership available to tax authorities would be a great step forward but, given the perception of opaque deal-making and capacity constraints, only when such registers were fully open to media, parliamentary, investor and public scrutiny would citizens be satisfied that the most egregious misrepresentation could be highlighted and tackled. Also registers would need to be automated to avoid snapshot information which could still conceal many deals. Cameron said that he would make such registers public if he could get international support to do so. At his first attempt he was unable to persuade his fellow leaders.

The EITI itself has taken a bold step into this field. The 2013 standard recommends that implementing countries maintain a publicly available register of the beneficial owners of the companies bidding, operating or investing in extractive assets.[3] Where these registers do not exist or are incomplete, the EITI recommends that implementing countries request companies to provide this information for inclusion in the EITI report. The EITI already requires that state-owned enterprises disclose their beneficial ownership in operating extractive companies. Pilot work is being undertaken to see if such registries could be a requirement for EITI implementing countries from 2015.[4]

The modern sociology of transparency determines that the burden is always on the holder of the data to justify why it should be confidential, rather than on the campaigner for transparency to justify why it should be open. As Cameron's experience shows, at first there might not be many countries willing to sign up, but for now the idea is out there and the transparency movement is difficult to resist. The EITI shows that companies and governments can agree some progress on this as it is in their mutual best interests. If Cameron and others continue to push, alongside international companies, a couple of high-profile G8 countries might agree to set a process in motion. Then, as the EITI shows, peer pressure around an idea can catch like wildfire until this transparency is the expectation, and opacity is the exception. The G20 in 2013 already agreed to begin automatically sharing information on financial flows between themselves by the end of 2015.

Of course, if governments agreed to publish beneficial ownership it might make bidding fairer, but these disclosures would not come close to preventing illicit financial flows or transfer mispricing. This would need to be a further step. Banks would need to disclose their flows. And this appears

---

3   Requirement 3.11 of the EITI Standard (EITI, 2013a).
4   EITI, 2015c.

to be an area in which governments are unable to legislate. If a government requires a bank to disclose, the bank has to choose between its customers, with whom it often has a confidentiality agreement, and its country of operation. Unless there are strong reasons to the contrary, they will side with the customer. However, some recent legislation to combat money laundering and terrorism is beginning to disturb even this sacred cow of accountholder secrecy.

Furthermore, there is surely a stronger case for opening up more information on bank accounts owned by governments in foreign countries (such as a government of Ghana account in the Cayman Islands). This should surely be public information and would no doubt reveal some odd findings about government holdings. The Bank of International Settlements has an oversight on these public bank accounts. As present, it is prevented from revealing this information by an ancient secrecy clause.

It would be audacious, but the authors think that there might be a model in which banks, activists and government representatives at the country level agree to a global standard for international flows into and out of banks based in the country—at first perhaps just covering public accounts. Above a certain threshold, the bank data would have to be attested by its auditor. Alongside a public register of beneficial ownership of companies registered in the country, this would give a picture of the inflows and outflows of the money of key individuals in a country. Spread across, for example, the G7/8 countries, and particularly the tax havens under their dependence, it could shine a huge light across global money tracking. Spread across the G20, and there would be little left hidden. Yet many G7/8 and G20 countries have established their bank sector on the principle of customer secrecy.

For this to work would require civil society activists at the table willing and able to monitor that full, comprehensive, timely and reliable data was being disclosed. More challengingly, it would also need banks to be held under pressure by a critical mass of governments of countries in which they want to operate, such that secretive banking no longer remains a viable business model. This business case for the banks to disclose does not yet exist and the pressure on the governments to demand it is too infant, but a shift towards the conditions for collective governance in this area might be sooner than many think. Given the enormous prize, though acknowledging the enormous challenges, the international community and civil society should be doing what they can to create the conditions for collective governance of transparent international financial flows as soon as possible.

11.2 Public infrastructure planning

Building public infrastructure is a notoriously challenging policy area. The Berlin Brandenburg Airport was originally planned to be opened in 2010, but encountered a series of delays and overspends due to poor planning, management and execution. It has still not been opened. In Edinburgh, the city tramway was due to be in service in early 2011, but several delays, disputes and cost overruns led to a three-year delay at more than three times the original cost. The 2014 Winter Olympics took place in Sochi, Russia. The bid estimated a cost of just under US$10 billion. This figure was many times more than any previous Winter Olympics and only slightly eclipsed for Summer Olympics by Athens and London. Due to poor planning and alleged corruption, the spending was estimated to be as much as US$50 billion.[5] These figures are, of course, small change compared with the missing billions spent on post-war reconstruction in Iraq. According to Transparency International, "Nowhere is corruption more ingrained than in the construction sector".[6]

Such planning disasters and corruption at the taxpayer's expense have brought down governments, suppressed growth and caused social and environmental upheaval. They have also led to increasing divestment from infrastructure by pension funds and sovereign wealth funds, and higher insurance costs. For the citizens, these investment disincentives are driving up the costs of food, transport and utilities, and in many countries this is driving a widening gap between the rich and the poor. Keynesian economics suggests that spending on public infrastructure—roads, transport, electricity, water, airports, pipelines, ports, communications, and so on—has multiple returns. History broadly bears out this theory, especially in developing countries. Given the potential for massive positive or negative economic, social and environmental outcomes, including grand corruption, there is clearly a strong interest from communities to engage in public debate and to monitor the activities of the companies and the government.

The Construction Sector Transparency Initiative (CoST) was established along the lines of the EITI model as a country-based collective governance approach for transparency and accountability in publicly financed construction, with the public provided with the information to hold government

---

5   Ormiston, 2013.
6   Construction was the focus of Transparency International's *Global Corruption Report 2005* (Transparency International, 2005).

and companies to account and to ensure better value for money in the construction sector. It aims "to reduce waste in public budgets, enables fairer competition in the private sector and increased opportunities for investors" (CoST, 2013). It claims to have identified improved efficiencies in the awarding of the contract, project design, procurement, quality, and time and budget overruns.

Despite these successes, unfortunately CoST has not made the same progress as the EITI and now faces serious financial constraints. The collective governance framework set out in this book might provide two explanations for this:

1. Is the business case strong enough? Despite public scandals such as the collapse of the Rana Plaza garment factory in Bangladesh that killed over a thousand people in 2013, there has not been the same anger about the macroeconomic and social impact of the sector at the global level as there is in the extractives sector. Alongside a dispersed industry with relatively low costs of entry, construction companies do not feel under the same investor pressure to get around the table.

2. The CoST model has sought to bring transparency to all aspects of the construction building chain at once, rather than working from a narrow point of consensus and broadening out to the meaningful. There is little doubt that there are problems of opacity and corruption all along the chain, from planning to contract to payments to maintenance, but an attempt to address it all before trust has been built is challenging.

The authors are supporters of CoST and sincerely hope that it overcomes these challenges and still becomes a success. Nonetheless, the authors would humbly suggest two shifts of focus for the model. First, from an assessment of the implementation of projects, where the issues are raw and threatening, to a shared planning model, which focuses on preventing poor decisions. Second, from national collective governance discussing often microeconomic projects, to subnational or urban collective governance where the conflict and impacts are more immediate and the companies fewer.

The approach of the Ecological Sequestration Trust is to bring together experts to create the open source forecasting models for city and regional infrastructure.[7] The models integrate economic, social and environmental

7   Further background to the model is described at http://ecosequestrust.org.

considerations to forecast the impact of infrastructure investments. The model is populated with land-use and ecological data from Earth observation satellites, and social and economic data from a combination of crowd sourcing and public records. Roads, railways and energy, water and communications networks can be built into the model by local owners and operators. It is envisaged that local universities will support model building and data verification with support from the trust. The model could then be used by a "Local Enterprise Partnership" of representatives from the local government, infrastructure construction firms and the local community. The model would provide the evidence base for their discussions and an agreed report could inform government and private-sector decisions on infrastructure spending. The model is being considered in the UK, Denmark, Africa and China.

Once the model succeeds in building trust, the Local Enterprise Partnership could be used to monitor the implementation of the project as envisaged by the CoST work.

11.3 International land acquisitions

Land acquisitions, especially by foreign companies, are a highly contentious area directly threatening the livelihoods of communities and citizens. There are increasing reports, for example, of agriculturally poor Gulf states buying up fertile land in Ethiopia, Sudan and Mozambique as future assurance against food insecurity, but leaving the land fallow in the meantime. The International Land Coalition of civil society, governments and international organizations reports nearly 40 million hectares acquired through over 1,000 such large-scale acquisition deals.[8] At the same time, farming communities around the land face enormous food shortages.[9]

There are also cases throughout the world of the speculative purchase of property which is then left unused, while the lack of affordable housing stock prices others out of the market and leaves many homeless. Financial cities, such as New York and London, appear to experience this behaviour the most paradoxically, with the financial crisis exacerbating the problem as the collapse of stock market bonds and shares pushes investors into the relatively safe property market. With the effects of the crisis hitting the poorest

---

8   See http://www.landmatrix.org/en/.
9   Harrigan, 2014.

worst, the gap for ordinary citizens between earning and owning widens, while non-productive speculation brings in vast returns.

These situations appear to the authors to contain all the basic preconditions for collective governance. These are both cases of government failure to penalize non-productive activity. Governments have failed to regulate since there is a need to attract foreign investment and they tend to lack the regulatory powers to restrict private purchase. Yet the behaviour of investors is threatening food security or affordable housing stock. In Ethiopia alone, around 35 million people are undernourished.

From the point of view of ordinary citizens, often they have been dispossessed and are locked out of owning or renting. In the case of land, they might even be prevented from providing for their own livelihood and sustenance. They could be using the unused land and even be provided with skills to improve their previous yields. This leads to increased anger and hostility.

From the point of view of investors, by sitting on unused land or property they are not optimizing their revenue. The large size of the acquisitions and either the weakness of government institutions to guarantee land rights in host countries, or the danger of a bursting speculative bubble, suggests that there is a significant risk for the buyer. Daewoo Logistics' contract to buy half the arable land in Madagascar not only contributed to bringing down a government, but was cancelled by the incoming regime in 2009. Therefore, investors might well be interested in a more secure, predictable and profitable governance arrangement.[10] Furthermore, these are often not very dispersed companies. Often the buyers are actually a few sovereign wealth funds looking for a diverse and strategic portfolio.

There appear to be surprisingly obvious win–win solutions available but these need facilitation of the various stakeholders—the government, the communities and the buyers. If a governance entrepreneur could bring these three groups together, they might be able to agree on a voluntary basis for some sort of affordable leasing arrangement either to individuals or to subnational levels of government. As a start, the International Land Coalition has developed a Land Matrix Global Observatory:[11] a tool that promotes transparency in land transactions and supports open data about land deals, from negotiation to implementation. In time, this tool could be overseen by semiformal and formal collective governance groups in the

10   Deininger and Byerlee, 2011.
11   See www.landmatrix.org.

affected countries, leading to conversations about the arrangements for the acquisitions of large land or property. If the buyers were not willing to take part in the discussions, then there could be the threat of mandatory legislation, perhaps with some international oversight, and a global campaign for more transparency of sovereign wealth funds.

11.4 Press regulation

There are few more contentious governance issues than press regulation. There are countless examples of media companies abusing their powers to distort public discourse and to pursue narrow commercial or ideological agendas. Often these commercial or ideological agendas can be internationally focused and consequently—as many have suggested about News International—with little interest in promoting informed domestic debate.

Media is a notoriously difficult sector to regulate because of the overlapping interests of the government and the companies. If a government tries to legislate or regulate too heavily, or if it takes control of too many media outlets, it too can be accused of determining public debate, controlling information and suppressing voices of opposition and investigation. There are also economic costs since the media brings important information to the market as well as being an economic sector in its own right.

Whatever the relationship between government and the media, ordinary citizens can feel excluded, exploited and misinformed. This concern is exacerbated by press scandals and intrusions or by government takeover of, or influence over, the media. Press self-regulation has had some success, but has not prevented all abuses, nor, more importantly, has it won the trust of the public. Civil society demands a safe platform in which to bring these "estates" of society to account.

The preconditions for collective governance appear to exist. There is clearly considerable public anger, an inability for government to regulate, dissatisfaction with attempts to self-regulate and, in most countries, a media sector dominated by a few major players with considerable investor and reputational risk. There is, of course, also a huge tail of small media outlets, blogs and tweets which would fall below the scope of collective governance.

A collective governance body could be brought together, at first, simply to develop agreed codes of behaviour. In due course, if sufficient trust can be established, the group might monitor complaints against the press and decide appropriate action, and even make suggestions on wider management of information.

Unsurprisingly, this quasi-statutory regulation was somewhere close to what was concluded in the UK by the Leveson enquiry into the culture, practices and ethics of the British press following a series of press scandals. Lord Leveson found that existing press self-regulation was not sufficient, and recommended a new independent multi-stakeholder body, which would have a range of sanctions available to it. Sign-up to the authority of the body by press would be voluntary, but incentivized by schemes such as a Kitemark standard, and an arbitration service would handle libel and breach of privacy claims. Non-participants in the scheme would, on the other hand, be subject to damages in cases brought against them.

It is an interesting model of collective governance which is still under formation and, if successful, could become a model for press regulation elsewhere.

12
Conclusions

The EITI journey has perhaps been a narrow experience of how govern-ance can operate beyond the usual boundaries of state institutions. There are many elements of the extractive industries that are somewhat untypical. However, it does highlight some important lessons that may be of value to other efforts as they set out in their journey "beyond governments", and to policy-makers, companies and academics. The authors do not claim any magic formula for good governance beyond governments, but their experi-ence does bring them to some clear conclusions.

Collective governance, such as the EITI, requires patience and enterprise, and an acceptance of an incremental and adaptive process. Successful implementation has to be built on a strong supporting network to help the progress over the challenges of communication without clearly defined objectives, of slow demonstration of impact, of linking with other efforts in the same field, and of keeping momentum for deeper implementation. The consensus among the vastly different stakeholders evolves, and so too must the aims and activities of the collective governance.

Collective governance is not a panacea for the problem of lack of politi-cal will, though it can be the best option available. The EITI has not solved the Angolan challenge whence it was born: how to manage natural resource transparently and accountably in that country. However, it has gone fur-ther than many could have imagined in difficult environments where it has been championed by skilled political reformers, such as in Nigeria, the Democratic Republic of Congo, Iraq, and so on. Such "beyond govern-ments" solutions to governance challenges themselves need to be carefully governed. They need to be designed and calibrated to build appropriate evolving institutions which can sustain demand for the process. This needs careful day-to-day management as well as sweeping strategic vision.

In the authors' view, there are few areas of governance in which collective governance can work and in which its benefits justify the efforts. Management of the extractive industries is one of the few examples, arising as it did from a failure of governments to address the concerns of citizens and a perfect storm of motivations for the right stakeholders to sit down at the same table. Sustaining that initial momentum has required strong leadership by all actors—government leaders willing to embrace reform and tackle inherent vested interests; company representatives willing to look beyond narrow self-interest to establish long-term licences to operate; and civil society representatives ready to couple activism with engagement. To repeat that formula in other sectors would undoubtedly take strong governance entrepreneurship.

Nonetheless, even though collective governance is tough and certainly not a simple solution, with some deep problems, it might be better than the alternatives. However hard collective governance might seem, it should be noted that intergovernmental processes such as the United Nations have struggled to address some of the biggest political issues and challenges of our time. Therefore there might be some policy areas that, though challenging, are better suited to collective governance solutions than any others. In order to be able to play their part, there needs to be a shift in mind-set from the international community—the development agencies, companies and non-governmental organizations.

Finally, the authors need to answer the question in the title of Chapter 1: "The irresistible rise of collective governance?" From their experience, the answer is that it is not as irresistible as some of the academic literature would have it. In most cases it is not desirable, but in an increasingly complex world of shifting balances of power between constituencies in society, there are certainly more opportunities. Even where there are opportunities, it will need a lot of governance entrepreneurship and skill to bring the efforts to a tangible result. To those wishing to embark on that journey, the authors offer their sincere good wishes and hope that they find this book of use.

Bibliography

Abrahamson, E. (2013). *Beyond Charity: A Century of Philanthropic Innovation*. New York, NY: Rockefeller Foundation.

Africa Progress Panel (2013). *Africa Progress Report—Equity in Extractives: Stewarding Africa's Natural Resources for All*. Retrieved from http://www.africaprogresspanel.org/wp-content/uploads/2013/08/2013_APR_Equity_in_Extractives_25062013_ENG_HR.pdf.

Annan, K. (2013). Foreword. In Africa Progress Panel, *Africa Progress Report—Equity in Extractives: Stewarding Africa's Natural Resources for All* (pp. 6-7). Retrieved from http://www.africaprogresspanel.org/wp-content/uploads/2013/08/2013_APR_Equity_in_Extractives_25062013_ENG_HR.pdf.

Ansell, C., & Gash, A. (2008). Collaborative governance in theory and practice. *Journal of Public Administration Research and Theory*, 18(4), 543-571.

Bannon, I.B., & Collier, P. (2003). *Natural Resources and Violent Conflict: Options and Actions*. Washington, DC: World Bank.

Barma, N., Kaiser, K., Minh, L., & Vinuela, L. (2012). *Rents to Riches: The Political Economy of Natural Resource-Led Development* (Washington, DC: World Bank).

Biermann, F., Chan, S., Mert, A., & Philipp, P. (2012). *Public–Private Partnerships for Sustainable Development: Emergence, Influence and Legitimacy*. Cheltenham, UK: Edward Elgar.

Boschini, A.B., Pettersson, J., & Roine, J. (2007). Resource curse or not: a question of appropriability. *Scandinavian Journal of Economics*, 109(3): 593-617.

Brandeis, L.D. (1914). *Other People's Money—and How Bankers Use it*. Chicago, IL: Cosmo Classics.

Bremmer, I. (2006). *The J Curve: A New Way to Understand Why Nations Rise and Fall*. New York, NY: Simon & Schuster.

Bronwen, M., 1999. *The Price of Oil: Corporate Responsibility and Human Rights Violations in Nigeria's Oil Producing Communities*. New York, NY: Human Rights Watch.

Browne, J. (2010). *Beyond Business: An Inspirational Memoir from a Remarkable Leader*. London, UK: Weidenfeld & Nicolson.

Carbonnier, G., Brugger, F., & Krause, J. (2011). Global and local policy responses to the resource trap. *Global Governance*, 17(2): 247-264.

CIA (Central Intelligence Agency) (2008). *The World Factbook*. Dulles, VA: Potomac Books.

Clarke, D. (2010). *Africa: Crude Continent—the Struggle for Africa's Oil Prize*. London, UK: Profile Books.

Collier, P. (2007). *The Bottom Billion: Why the Poorest Countries Are Failing and What Can Be Done About it*. New York, NY: Oxford University Press.

Collier, P., & Hoeffler, A. (1998). On economic causes of civil war. *Oxford Economic Papers*, 50(4): 563-573.

CoST (Construction Sector Transparency Initiative) (2013). What is CoST? Retrieved from http://www.constructiontransparency.org/the-initiative?forumboardid=1&forumtopicid=1.

Cramer, A. (2013, January 23). Solving problems the civil society way. *BSR: The Business of a Better World*. Retrieved from http://www.bsr.org/en/our-insights/blog-view/solving-problems-the-civil-society-way.

David-Barrett, L., & Okamura, K. (2012). *The Transparency Paradox: Why Do Corrupt Countries Join EITI?* Berlin, Germany: European Research Centre for Anti-Corruption and State-Building.

Deininger, K., & Byerlee, D. (2011). *Rising Global Interest in Farmland: Can it Yield Sustainable and Equitable Benefits?* Washington, DC: World Bank:

DFID (Department for International Development) (2003). *Report of the Extractive Industries Transparency Initiative: London Conference, 17 June 2003*. London, UK: Author:

Donahue, J., & Zeckhauser, R. (2011). *Collaborative Governance: Private Roles for Public Goals in Turbulent Times*. Princeton, NJ: Princeton University Press.

EITI (Extractive Industries Transparency Initiative) (2003a). The EITI Principles. Retrieved from http://eiti.org/eiti/principles.

—— (2003b). Investors' statement on transparency in the extractives sector. Retrieved from http://eiti.org/investors-statement-transparency-extractives-sector.

—— (2005). The EITI Criteria. Retrieved from http://eiti.org/eiti/criteria.

—— (2006). *EITI Validation Guide*. Retrieved from https://eiti.org/files/document/validationguide.pdf.

—— (2009). EITI Rules. Retrieved from https://eiti.org/files/2010-02-24%20EITI%20Rules_0.pdf.

—— (2013a). The EITI Standard. Retrieved from http://eiti.org/document/standard.

—— (2013b). EITI and public administration efficiency. Retrieved from https://eiti.org/blog/eiti-and-public-administration-efficiency.

—— (2013c). *Articles of Association for the Extractive Industries Transparency Initiative (EITI)*. Retrieved from https://eiti.org/files/EITI_ArticlesofAssociation_EN.pdf.

—— (2014). *EITI Association Code of Conduct*. Retrieved from http://eiti.org/document/code-of-conduct.

—— (2015a). *Civil Society Protocol*. Retrieved from https://eiti.org/files/EITI_CivilSocietyProtocol_Jan2015.pdf.

—— (2015b). Companies. Retrieved from https://eiti.org/supporters/companies.

—— (2015c). Pilot project: beneficial ownership. Retrieved from http://eiti.org/pilot-project-beneficial-ownership.

Forbes, K. (2013, September 7). Ugly reality of tax dodgers of Africa. *Africa Review*. Retrieved from http://www.africareview.com/Special-Reports/Ugly-reality-of-the-tax-dodgers-of-Africa/-/979182/1983112/-/gsktr6z/-/index.html.

Ghazvinian, J. (2008). *Untapped: The Scramble for Africa's Oil*. Orlando, FL: Houghton Mifflin Harcourt Publishing.

Global Witness (1999). *A Crude Awakening: The Role of the Oil and Banking Industries in Angola's Civil War and the Plunder of State Assets*. Retrieved from http://www.globalwitness.org/sites/default/files/pdfs/A%20Crude%20Awakening.pdf.

Harrigan, J. (2014). *The Political Economy of Food Sovereignty in the Arab World*. London, UK: Palgrave Macmillan.

Haufler, V. (2010). Disclosure as governance: the Extractive Industries Transparency Initiative and resource management in the developing world. *Global Environmental Politics*, 10(3): 53-73.

Herbert, D.D. (2004). *They Are Lincoln: Abraham Lincoln and his Friends*. New York, NY: Simon & Schuster.

Humphreys, M., Sachs, J., & Stiglitz, J. (2007). *Escaping the Resource Curse*. New York, NY: Columbia University Press.

International Civil Society Centre (2014). *Multi-Stakeholder Partnerships: Building Blocks for Success*. Berlin, Germany: Author.

IMF (International Monetary Fund) (2014). Reports on the observance of standards and codes (ROSCs). Retrieved from http://www.imf.org/external/NP/rosc/rosc.aspx.

Jarvis, M. (2013, May 22). Michael Jarvis: The multi-stakeholder initiatives mid-life crisis: time for a support group? *Business Fights Poverty*. Retrieved from http://community.businessfightspoverty.org/profiles/blogs/michael-jarvis-the-msi-mid-life-crisis-time-for-a-support-group.

Karl, T.L. (1997). *The Paradox of Plenty: Oil Booms and Petro-States*. San Francisco, CA: University of California Press.

LeBillion, P. (2003). Buying peace or fuelling war: the role of corruption in armed conflicts. *Journal of International Development*, 15(4): 413-426.

Levy, B. (2014). *Working with the Grain: Integrating Governance and Growth in Development Strategies*. Oxford, UK: Oxford University Press.

Litvin, D. (2013). *Empires of Profit: Commerce, Conquest and Corporate Responsibility*. Knutsford, UK: Texere Publishing.

Maas, P. (2009). *Crude World: The Violent Twilight of Oil*. New York, NY: Random House, Inc.

Margonelli, L. (2007). *Oil on the Brain: Adventures from the Pump to the Pipeline*. New York, NY: Random House.

Mantovani, M. (2012). The business case for collective action. In M. Peith (Ed.), *Collective Action: Innovative Strategies to Prevent Corruption* (pp. 73-82). Zurich, Switzerland: Dike.

Marianne, B., & Liese, A. (2014). *Transitional Partnerships: Effectively Providing for Sustainable Development?* New York, NY: Palgrave Macmillan.

Moody-Stuart, M. (2014). *Responsible Leadership: Lessons from the Front Line of Sustainability and Ethics*. Sheffield, UK: Greenleaf Publishing.

Morrison, J., & Wilde, L. (2007). *The Effectiveness of Multi-Stakeholder Initiatives in the Oil and Gas Sector*. Frome, UK: TwentyFifty Ltd.

Mulholland, J. (2011, February 20). Bono on Africa: "What excites me is thinking about its future". *The Guardian*. Retrieved from http://www.theguardian.com/commentisfree/2011/feb/20/bono-africa-economic-growth.

Natural Resource Governance Institute (2006). Gabon crackdown on civil society groups prompts swift outcry from Publish What You Pay US coalition. Retrieved from http://www.resourcegovernance.org/news/press_releases/gabon-crackdown-civil-society-groups-prompts-swift-outcry-publish-what-you-pay-u.

—— (2008). Congo-Brazzaville: legal defence funding for PWYP activists. Retrieved from http://www.resourcegovernance.org/grants/congo-brazzaville-legal-defense-funding-pwyp-activists.

—— (2009). Harassment of activists stalls EITI process in Niger. Retrieved from http://www.resourcegovernance.org/news/harassment-activists-stalls-eiti-process-niger.

—— (2014). The Resource Governance Index (RGI) measures the quality of governance in the oil, gas and mining sectors of 58 countries. Retrieved from http://www.resourcegovernance.org/rgi.

—— (2015). Grants. Retrieved from http://www.resourcegovernance.org/grants/.

NEITI (Nigeria Extractive Industries Transparency Initiative) (2012). *10 Years of NEITI Reports: What Have We Learnt?* Retrieved from http://www.neiti.org.ng/sites/default/files/publications/uploads/ten-years-neiti-reports.pdf.

Olson, M. (1965). *The Logic of Collective Action: Public Goods and the Theory of Groups.* Cambridge, UK: Harvard University Press.

Open Government Partnership (2015). What is the Open Government Partnership? Retrieved from http://www.opengovpartnership.org/about.

OpenOil (2012). *Exploring Oil Data: A Reporter's Handbook.* Retrieved from http://openoil.net/exploring-oil-data/.

Opoku-Mensah, P.Y. (2007). The state of civil society in sub-Saharan Africa. In V.F. Heinrich & L. Fioramonti (Eds.), *CIVICUS Global Survey of the State of Civil Society, Volume 2: Comparative Perspectives* (pp. 75-90). Bloomfield, CT: Kumarian Press.

Ormiston, S. (2013, February 4). Susan Ormiston: Russia's $50b Sochi Olympics gamble. *CBC Sports.* Retrieved from http://www.cbc.ca/sports/olympics/susan-ormiston-russia-s-50b-sochi-olympics-gamble-1.1409726.

O'Sullivan, D. (2013). *What's the Point of Transparency?* Oslo, Norway: Open Society Fellowship:

OUP (Oxford University Press) (2002). *Concise Oxford English Dictionary.* New York, NY: Author.

Prakash, A. (2001). Why do firms adopt 'beyond-compliance' environmental policies? *Business Strategy and The Environment,* 10(5), 286-299.

Richter, A. (2010). Prepared statement of Anthony Richter, Chairman of the Governing Board, Revenue Watch Institute. In *The Link Between Revenue Transparency and Human Rights: Hearing Before the Commission on Security and Cooperation in Europe, Washington, DC, April 22, 2010* (pp. 42-50). Retrieved from http://www.csce.gov/index.cfm?FuseAction=Files.Download&FileStore_id=2297.

Roberts, A. (2006) *The Wonga Coup: Guns, Thugs and a Ruthless Determination to Create Mayhem in an Oil-Rich Corner of Africa.* Cambridge, UK: Profile Books.

Ross, M. (2001). *Extractive Sectors and the Poor.* Boston, MA: Oxfam:

—— (2013). EITI 2.0. Retrieved from http://goxi.org/profiles/blogs/eiti-2-0?xg_source=msg_mes_network.

Ruggie, J. (2007). *Business and Human Rights: Mapping International Standards of Responsibility and Accountability for Corporate Acts. Report of the Special Representative of the UN Secretary-General on the Issue of Human Rights and Transnational Corporations and Other Business Enterprises.* New York, NY: UN General Assembly.

Scanteam (2011). *Achievements and Strategic Options: Evaluation of the Extractive Industries Transparency Initiative.* Retrieved from http://eiti.org/files/2011-EITI-evaluation-report.pdf.

Schmaljohann, M. (2013). Transparency Commitments as a Tool for Fighting Corruption? Evidence from the Extractive Industries Transparency Initiative. Retrieved from https://eiti.org/files/Schmaljohann_2013_dp538.pdf.

Scovel, J. (1898). Thaddeus Stevens. *Lippincott's Monthly Magazine,* April, 548-550.

Shaxson, N. (2008). *Poisoned Wells: The Dirty Politics of African Oil.* New York, NY: Palgrave Macmillan.

Transparency International (2005). *Global Corruption Report 2005: Corruption in Construction and Post-Conflict Reconstruction.* Berlin, Germany: Author.

Tax Justice Network (2009). *Where on Earth Are You? Major Corporations and Tax Havens.* Retrieved from http://www.taxjustice.net/cms/upload/pdf/45940CCBd01.pdf.

Tax Justice Network Africa (2011), *Tax Us If You Can: Why Africa Should Stand Up for Tax Justice.* Cape Town, South Africa: Pambazuka Press.

Van Oranje, M., & Parham, H. (2009). *Publish What We Learned: An Assessment of the Publish What You Pay Coalition.* London, UK: Publish What You Pay:

Warner, M., and Sullivan, R. (2004). *Putting Partnerships to Work: Strategic Alliances for Development Between Government, the Private Sector and Civil Society* (Sheffield, UK: Greenleaf Publishing).

Wintour, P. (2013, June 16). David Cameron urged to extend tax reporting plan. *The Guardian*. Retrieved from http://www.theguardian.com/politics/2013/jun/16/david-cameron-extends-tax-plan.

World Bank (2011). *CPIA 2011 Criteria*. Retrieved from http://www.worldbank.org/ida/papers/CPIAcriteria2011final.pdf.

World Bank Development Grant Facility (2007). DGF recipients summary table. Retrieved from http://siteresources.worldbank.org/INTEXTINDTRAINI/Resources/dgf_summary.pdf.

About the authors

Eddie Rich has worked in development for over 20 years. In 1996–98 he was the DFID representative to Angola. In charge of a medium-sized aid programme, it soon became clear to him that the development debate needed to move well beyond aid. This was reinforced when he was part of the small team working on the UK government's ground-breaking White Paper on international development, *Eliminating World Poverty: Making Globalisation Work for the*  *Poor* (2000). The paper advocated a more coherent approach to international development policy highlighting issues of trade, governance, security, codes and standards, finance and the environment. He also co-authored a background study for this on poorly performing states. He was head of DFID's corporate social responsibility team when the nascent Publish What You Pay coalition came to DFID with an idea for a transparency initiative in the extractive sector in 2001. He continued to work on the Extractive Industries Transparency Initiative in those infant times until 2004 when he moved to Kenya as deputy head of DFID Kenya. When Kenya started exploring for deep sea oil in 2006/7, he was thrown back into the EITI debate. Since 2007, Eddie has been the deputy head of the EITI International Secretariat where he has led on Africa and the Middle East as well as oversight for finance, human resources, communications and the global conferences.

Jonas Moberg began his career as a Swedish diplomat, serving in Maputo and London. With a growing interest in the crossroads of governments, private sector and civil society, in he 2002 joined the Prince of Wales International Business Leaders Forum (IBLF). Having been the director of the IBLF's "Business and Human Rights" and "Business and Corruption" programmes, he became a senior adviser to the UN Global Compact in 2005. Since 2007, Jonas has been the head of the EITI International Secretariat.